This book challenges several widespread views concerning Aristotle's methods and practices of scientific and philosophical research. Where does the boundary between the animal and the plant kingdom come? How can supposed cases of spontaneous generation and metamorphosis be explained? What do the processes of reproduction, growth and digestion have in common that allows Aristotle to apply his theory of 'concoction' to each? What contributions to astronomical theory does Aristotle hope to make? Does he claim to apply precisely the same explanatory schema to every mode of sense-perception? How does he use his notion of 'nature' in his politics? Does he have unified concepts of analogy, of metaphor, of demonstration? Professor Lloyd explores generally unrecognised tensions between Aristotle's deeply held *a priori* convictions and his remarkable empirical honesty in the face of complexities in the data or perceived difficult or exceptional cases. The picture that emerges of Aristotle's actual engagement in scientific research and of his own reflections on that research is substantially more complex than is usually allowed.

Aristotelian explorations

Aristotelian explorations

G. E. R. Lloyd

*Professor of Ancient Philosophy and Science in the
University of Cambridge, and Master of Darwin College*

CAMBRIDGE
UNIVERSITY PRESS

University Printing House, Cambridge CB2 8BS, United Kingdom

Cambridge University Press is part of the University of Cambridge.

It furthers the University's mission by disseminating knowledge in the pursuit of education, learning and research at the highest international levels of excellence.

www.cambridge.org
Information on this title: www.cambridge.org/9780521554220

© Cambridge University Press 1996

This publication is in copyright. Subject to statutory exception and to the provisions of relevant collective licensing agreements, no reproduction of any part may take place without the written permission of Cambridge University Press.

First published 1996
First paperback edition 1999

A catalogue record for this publication is available from the British Library

ISBN 978-0-521-55422-0 Hardback
ISBN 978-0-521-55619-4 Paperback

Cambridge University Press has no responsibility for the persistence or accuracy of URLs for external or third-party internet websites referred to in this publication, and does not guarantee that any content on such websites is, or will remain, accurate or appropriate.

Contents

Preface		*page* ix
	Introduction: reading Aristotle	1
1	The theories and practices of demonstration	7
2	The relationship of psychology to zoology	38
3	Fuzzy natures?	67
4	The master cook	83
5	Spontaneous generation and metamorphosis	104
6	The varieties of perception	126
7	The unity of analogy	138
8	Heavenly aberrations: Aristotle the amateur astronomer	160
9	The idea of nature in the *Politics*	184
10	The metaphors of *metaphora*	205
Bibliography		223
Index of passages referred to		231
General index		236

Preface

This book originates in lectures and seminars that I have given over the last eight years, in Cambridge, Oxford, Paris, Boston and Princeton. Two of the chapters (the first and the second) have been published before in English and one (chapter 9) in French. Their provenance, and my policies for revising them here, are explained in notes to the chapters concerned. Other studies stem from presentations to Dr David Charles' workshop in Oxford in May 1994 (chapter 3), to a Laurence reading week in Cambridge in June 1994 (chapter 8), to a conference organised by Professor Barbara Cassin in Paris in October of the same year (chapter 5), or derive from my lecture courses and seminars in Cambridge over the last four years. Since they all deal with the broad problem of the interactions between Aristotle's *theories* of scientific and philosophical research and his actual *practices* in a variety of domains, I have thought it worthwhile to collect and revise them for publication as a set of studies devoted to that subject.

My thanks go in the first instance to my lecture and seminar audiences, to all those who have participated in the discussion on the various occasions when this material has been presented. I have corresponded on several controversial questions with a number of scholars who have generously offered helpful comments on the issues, and especially with Professors Allan Gotthelf, Wolfgang Kullmann, Jim Lennox and Pierre Pellegrin. Finally I have been most fortunate in receiving detailed criticisms of a draft of the entire text from both Professor Myles Burnyeat and Dr David Charles. This has caused me to qualify, elaborate, and, I trust, improve, my interpretation in innumerable places. None of those I have mentioned should be presumed to agree with many, or indeed any, of the points of view here expressed. But I must record my gratitude to them all for the help they have given me in clarifying my arguments on complex and difficult questions.

Introduction: reading Aristotle

When, nearly thirty years ago, I wrote a general introduction to Aristotle, developmental studies were all the rage. Stimulated by Jaeger's pioneering work, many Aristotelian scholars attempted to plot the changes in Aristotle's views, either in particular areas of his thought, or overall.

Much of this work was naively conceived and poorly executed. Scholars saw inconsistencies between statements where reconciliation was not just possible, but obvious. A recurrent flaw was that the changes of heart that Aristotle was supposed to have undergone were left unexplained: scholars were content to suggest *that* his views had altered, but did not offer reasons *why* they should have done so. I have had occasion myself to modify some of my own overenthusiastic early developmental hypotheses (Lloyd 1991, pp. 1ff.).

Yet developmental studies did have one great strength, compared with the tendency to systematisation that had been the previous orthodoxy. They did not treat the thought of Aristotle as a monolith. They allowed for the possibility that not everything in the Corpus had been fully integrated into a single definitive whole.

The fashion for developmental studies lapsed some time ago. In another area of academic study, meanwhile, in literature, the entire tradition of attempting to arrive at fixed interpretations of writers, past or present, came under blistering, deconstructive attack. Such interpretations merely reflected their authors' own hidden agenda and prejudices, and in pursuing the will-o'-the-wisp of closure, blocked, rather than stimulated, study.

This fashion, too, had its excesses and has since waned, leaving very little mark on the work of Aristotelian specialists. Yet they, whether impressed by deconstructionist views or not, certainly need to pay due attention to the cautions emanating from literary and historical studies. This is especially necessary as an antidote to the tendency to treat Aristotle as a contemporary philosophical colleague. That tendency is, no doubt, often driven by the entirely laudable desire not just to take his work seriously, but to learn from it. But that has to be done without assuming that he is one of us (whoever 'we' are) or that his thought is somehow disembodied and timeless.

The nature of the evidence with which we have to deal poses special problems of reading and interpretation over and above those that affect *all* texts. What has survived reflects the vagaries of transmission and the choices of the transmitters. None of the literary works of Aristotle is extant in its entirety. If they too had survived, we cannot say *what* difference that would have made, but very probably it would have been considerable. The treatises, for their part, have been subject to editorial intervention of various types – ranging from the grouping of related discussions together, through the addition of cross-references or bridging material, to sheer interpolation. However, much of this editorialising, of the first two kinds, may well have been by Aristotle himself.

It is clear that these are not works prepared for publication as literature, even by the criteria for 'publication' that apply to the ancient Greek world. But we can go further. The existence of doublets, and of references to the conclusions of earlier discussions that do not correspond to the contents of the discussions we have, casts a long shadow over the issue of the extent to which the treatises represent *finished* work, that is that Aristotle considered such.

Of course the problem of doublets – the repetition of closely similar material or different versions of the same text in different manuscript traditions, as in the case of *Physics* VII 1–3 – can, in principle, always be resolved by arguing that one or other treatment is inauthentic, the work of a pupil, maybe, attempting a précis of the master's lecture. That is the line of argument that has often been adopted with regard to the zoological treatises and is, in the final analysis, the preferred solution of Wardy's careful study of *Physics* VII (Wardy 1990). On the other hand no one believes that one or other version of the critique of Plato in *Metaphysics* A and M is inauthentic. Moreover when in the opening chapter of *Metaphysics* H the conclusions of a preceding discussion are summarised, this tallies well enough with the problems tackled in *Metaphysics* Z, but seems to discount some of the results arrived at in the book as we have it (cf. Woods et al. 1984). Again *Metaphysics* Z 7–9 has the marks of an independent self-contained inquiry, which was subsequently incorporated into the text we have. All of this suggests, what is in any case plausible enough, that on certain topics Aristotle's writing went through several drafts, as we might call them, and that in turn sounds that general cautionary note that I have mentioned about how definitive the version we have was.

Plato alerts his readers to the need to be context-, and in his case also character-specific, in reading every sentence in his dialogues. Every statement is made by a particular speaker to a particular audience at a particular juncture in the course of what is represented as a live conversation, whether told in *oratio recta* or reported at one or more removes in *oratio obliqua*. But similarly, if not quite equally, the statements we find in Aristotelian texts have always to be taken as embedded in their, more or less richly crafted, contexts.

Plato interposes the dialogue, like a veil, between himself and us, the reader. But Aristotle too demands, though less obviously, similar hermeneutic skills, and not just for reasons connected with the abstractness and complexity of his thought.

The starting-point of this set of studies is a sense of the difficulty of the work that Aristotle undertook. I do not mean thereby to imply that we necessarily have any clear idea of how he saw that 'work'. I mean something far more basic. The list of his interests is a daunting one, even if we just use the familiar, if potentially misleading, labels we attach to certain subject-areas: logic, physics, psychology, zoology, metaphysics, ethics, politics, rhetoric, poetics.

In most of these fields he is extraordinarily innovative. Of course his own predecessors and contemporaries, with whom he is in constant debate, had opened up many fundamental philosophical and scientific questions; but Aristotle's own researches add many more items to his agenda. Among those we shall be discussing in the chapters that follow are: how is any given animal to be defined – that is, how is definition to be applied in zoology? Where does the boundary between plants and animals come? How are such processes as digestion and reproduction to be explained? How are apparent instances of spontaneous generation and of metamorphosis to be accounted for? Do all five modes of perception conform to a single analytic schema? How are the varieties in the movements of the heavenly bodies to be explained? How far does the concept of the natural apply also to the political domain? What are the limits of the usefulness of *metaphora*, transfer?

These problems relate only to a handful of the areas in which Aristotle worked. Nor could they be claimed to be a representative selection of fundamental issues (if such were possible). But they give some indication of the difficulty of the tasks he set himself: indeed the nature of the possible solutions available to him (and in some cases still also to us) leaves a lot to be desired. Of course, what appears readily soluble is heavily influenced by historical determinants. But Aristotle himself registers the difficulty he encounters in many of the cases we shall be considering.

Does he, for those tasks, have a single, all-embracing methodology, or at least one for each type of task? Is he consistent in his application of the approved methodology? Is he consistent in his aims, indeed, and in the criteria he adopts for a satisfactory explanation or resolution of the problems? Is he confident in the results he proposes, whether or not the results stem from the use of the methodology?

There are many indications, in his texts, that point towards the ideal of an affirmative answer to those questions, at least over a range of issues, and no one can doubt the extraordinary breadth of the field of application of such cross-disciplinary concepts as form and matter, potentiality and actuality, order and nature itself. Yet there are also plenty of signs, if we do not discount them, of tentativeness, of hesitation, of the pluralism and open-endedness of his approach, of a readiness to backtrack, to qualify and modify even fundamental doctrines and principles.

Of course many would be tempted, many have been, if not to discount those signs, at least to minimise their significance, in the name of the coherence and unity of Aristotle's thought. There is no way in which *a priori* the rights and wrongs of *that* methodological principle can be settled as against what is adopted here. But this exploration is committed to allowing the possibility that Aristotle's explorations indeed ran certain risks.

This is not to say that he was simply unmethodical, merely opportunistic in applying or not applying his overarching methodological concepts or general theories. Not at all. It is rather that his methodology is far more pluralist than is often allowed, and in particular that it is responsive to the need for adaptation in the light of the demands and circumstances of different problem areas, responsive, for instance, to the state of research in the field and to its difficulty.

The first detailed case-study tackles one of his most powerful methodological concepts of all, that of demonstration, and argues – against the grain of much synthesising scholarship devoted to showing the unity of Aristotle's thought – not just that there are several different modes of demonstration, both in theory and in practice, in different areas of Aristotle's work, but also that he recognises this plurality. The subsequent studies investigate other aspects of his theories and practices, endeavouring to use a sustained analysis of the practices to illuminate the theories as well as re-examining the theories to help interpret the practices.

The picture that will emerge is not, nor should we expect it to be, a uniform one. The extent to which Aristotle's general theories are at risk in the face of the exceptions he is ready to admit varies – depending on the general theory and the nature of the exceptions. The extent of his own first-hand engagement in research also varies (as the example from astronomy, in chapter 8, will bring out) and this has important repercussions on the status of the methodological recommendations he makes. But if he sometimes offers methodological pronouncements just for their own sake – at the limit, purely for form – that highlights the contrast with those instances where the methodology has work to do, as providing the theory that guides the practice, or rather, as we shall repeatedly see, the theories that guide the practices.

This book ascribes risk-taking to Aristotle and in doing so itself runs certain risks, for to argue for the unevenness in the grain of an author's work is always more difficult that to do so for its homogeneity, especially when the author has so often been held up as a paragon of systematisation. Yet that view of Aristotle has contributed not a little to the neglect of aspects of the difficulty of the work he was doing – and of his recognition of that difficulty. Besides, since the thesis of the book is that those risks are a price Aristotle paid for deep involvement in first-hand research of considerable complexity, it is only natural for a similar price to have to be paid for engaging to the full in the difficulties of his interpretation. Yet the book also argues that Aristotle's conclusions are less definitive than they are often represented to be: and in this too it mirrors itself, for there can certainly be no question of certainty about the issues that we are about to explore.

CHAPTER I

The theories and practices of demonstration*

The question of the relationship between Aristotle's theory of science set out in the *Posterior Analytics* and his actual practice in the physical treatises is, it may be felt, a hoary old chestnut indeed.[1] Faced with a series of apparent discrepancies between the two, starting with the notorious mismatch between the attention devoted to the theory of the syllogism in the *Organon* and the evident lack of good-looking actual examples of syllogisms in the physical works, what are we to say? Crudely, one may distinguish two main schools of interpretation: I shall call them lumpers and splitters, the lumpers all for reconciliation, the splitters insisting on the differences and offering various suggestions as to how they may have come about. One such suggestion is that the *APo.* is not about doing science, but about teaching it. Another, allied to that, is that it is about presenting the results of a finished or at least a mature science. On either view that means that the physical treatises do not need to conform to the schemata set out in *APo*. In places the physical treatises may be provisional, preliminary, exploratory studies, and some would say that their method would be better described as dialectical than as demonstrative. A third more drastic, some may even think desperate, splitter-type theory invokes the now unfashionable style of developmental hypothesis, though

---
* This article is reprinted in its original form from the *Boston Area Colloquium in Ancient Philosophy* VI (1990), edd. J. J. Cleary and D. C. Shartin (Lanham, 1992), pp. 371-401, with the correction of misprints and the removal of remarks relating to my oral presentation at Boston University in April 1990. But I have added some additional notes identified by an asterisk, and a postscript.
[1] Among the most important contributions to the debate on this issue in recent years have been those of Barnes 1969/1975 and 1981, Bolton 1987, Düring 1961, Furth 1988, Gotthelf 1987, Kullmann 1974, Lear 1980, Lennox 1987a, Leszl 1981, Mignucci 1975, Rossitto 1984, Smith 1982, Sorabji 1980, Waterlow 1982, Wieland 1975 – and most recently Wians 1989 (an article I had not seen when this chapter was first produced).

it would take a desperado indeed to suggest either that the *APo.* antedated or that it postdated the *entire* corpus of physical treatises, that is everything from the *Physics* to the *de Generatione Animalium* in the Berlin edition. However even on a more moderate and more plausible developmental view there is still a problem in specifying *why* Aristotle should have changed his mind, in one direction or another, if and when he did.

But against the splitters, reconciling lumpers have recently been on the march, minimising or even cancelling the apparent discrepancies and going on the attack and insisting that in all sorts of ways that have been underestimated or ignored in the past the physical treatises *do* conform to the schemata presented in the *APo.* A fair number of pieces collected in the volume recently edited by Gotthelf and Lennox (1987) adopt such a stance, as Bolton does, for instance, in his comparison between the theory of definition in *APo.* and Aristotle's actual discussion of seed in *GA* I.[2] Lennox, too, in his analysis of the types of explanation actually given in the zoological treatises, concluded that they conform quite closely to those advocated in *APo.* Thus where *APo.* distinguishes between incidental and unqualified understanding, between what Lennox calls type A explanations that explain certain facts but *not* through the correct, primary explanantia, and type B explanations that do just that, so we can exemplify both types in the zoological works.[3] Gotthelf in turn[4] describes himself as facing 'head-on' the question of whether the *Parts of Animals* exemplifies anything approaching an axiomatic structure: are there first principles at the base of the complex explanatory structure? And he answers yes to both questions, though not without some difficulty. Noting that there is nothing *prima facie* axiomatic about the structure of *PA*, he has some softer formulations of his position – the structure is 'broadly axiomatic' or it is 'axiomatic-like' (in being linear).[5] As for the first principles he detects in the physical treatises, that seems to me – as I shall argue later – to trade rather on the ambiguity of the term 'first principles': but more on that in due course. Meanwhile Gotthelf also lends his support[6] to a suggestion of Aryeh Kosman's

[2] Bolton 1987.
[3] Lennox 1987a, pp. 93ff.
[4] Gotthelf and Lennox 1987, p. 68.
[5] Gotthelf 1987, pp. 175, 179, 186.
[6] Ibid. p. 195.

Theories and practices of demonstration

to the effect that *APo.* would be better seen *not* as requiring that proper science *be* formal, but simply as offering a formal description of proper science – though that still means (as Gotthelf acknowledges) that the *APo.* theory requires of the natural philosopher that his exposition be *puttable* into the appropriate form, even if not that it actually be so put. While I agree entirely with Gotthelf on the importance of the biology for understanding Aristotle, we disagree on some of the lessons to be learnt.

Now one common feature that runs through much of the recent discussion is that the problem of the relationship between the *Organon* and the physical treatises is construed holistically. Certainly so far as the theory and practice of *apodeixis* – demonstration – go, the controversy is sometimes conducted as if there were, in effect, *just the one* theory and as if one should expect to be able to generalise satisfactorily about *the* practice. Actually much of the debate is a good deal more ambitious than that and takes it that what we have to compare and contrast, and try – or fail – to reconcile, is nothing less than the theory of *science* and its practice, that is to say including, not just demonstration, but also definition, the status and method of discovery of the starting-points generally, the modes of deductive reasoning to be used in explanation, the modes of explanation themselves and much else besides. Of course it would be nice to think that Aristotle's thinking on all these topics was always and everywhere entirely uniform, consistent, unified, and systematic, allowing one to effect a grand reconciliation between the theory as a whole and the practice as a whole. Yet in reality the situation may be – and in the case of demonstration I shall argue that it is – appreciably richer. Oversystematising Aristotle should be avoided, where there are signs of a certain tentativeness and flexibility in both theory and practice. Some of the unresolved areas in current scholarly debate may – let us hope – reach some clarification and resolution: but we have to allow the possibility – the splitter in me insists – that the grand synthesis and reconciliation of the Aristotelian theory and practice will fail for reasons that lie in the Aristotelian texts themselves, and not just because of modern scholarly shortsightedness on one side of the debate or the other.

So to the topic of demonstration in particular, and I use that throughout as a conventional translation for Aristotle's *apodeixis* and I shall concentrate on occasions when this term and the cognate verb *apodeiknumi* are in play in the Aristotelian texts themselves, leaving to one side further issues that could only be resolved by spreading the net wider to include the whole range of other *deik*-forms and compounds, let alone other vocabulary used in one context or another to signify showing, manifesting, proving, demonstrating. It is agreed on all sides that *apodeixis* is a key concept in the *Organon* and the question of how far, in this regard, the physical treatises practise what the *Organon* preaches has (as I said) been pursued with some energy in recent studies. However the very variety of contexts in which this term is deployed has not always received the attention it deserves: indeed it has often been all but totally ignored. My aim in this paper is first to broaden the scope of the data to be taken into consideration, to see what light that throws on the general problem. To take stock of Aristotle's use of *apodeixis* we cannot limit ourselves to the *Organon* and the physical treatises, but have also to take into account evidence from, for example, the *Rhetoric*, the moral philosophy, the *Metaphysics*. Broadly speaking, my argument will be that when we do that, we have every reason to be wary of talking of *the* theory and *the* practice of demonstration, as if there were just the one of each. On the contrary we have good evidence that there are *several* theories and indeed several practices, and not just more, and less, strict concepts of *apodeixis* but *apodeixeis* that vary with certain features of the subject-matter under consideration and with the nature of the inquiry undertaken. That may be blindingly obvious (I hope it will be): but the implications for the strategic problems of the relationship between the *Analytics* and the physical works have often been missed.

But I must begin at the beginning with a brief reminder of what the *Analytics* have to say on the topic. *APo.* 71b17f. of course defines *apodeixis* as *sullogismos epistēmonikos*, though neither term in the definiens is exactly totally pellucid. Thus by *sullogismos* he sometimes means (as is well known) no more than deductive reasoning. Those who see a pre-syllogistic phase in Aristotle's theory of demonstration lying behind the *APo.* as we have it would no doubt want to leave *open* the general interpretation and not immediately foreclose with the translation 'syllogism'.[7] But whatever may or may not be true of what lies behind the *APo.*, the *Analytics* as we have

---

[7] Cf. e.g. Barnes 1981.

them – both *APr.* and *APo.* – use the developed theory of the syllogism that is set out in *APr*. That theory is not just a minor excrescence on the surface of the discussion in *APo.*, a skin that can easily be peeled off leaving the discussion intact. The theory informs the discussion through and through. Thus not just in his opening chapters on demonstration itself, but later, throughout his analysis of such topics as error, the limits of proof, the provability of definition and so on, he *generally* operates with the theory of the syllogism. That is clear not just from the examples given but also from his repeated references to the three figures and to results obtained in the analysis of syllogism in *APr*. Wherever Aristotle's thinking may have come from before he composed the *APo.*, in the treatise we have demonstration is generally ineluctably syllogistic in form.

The second term in the definition at *APo.* 71b17f., *epistēmonikon*, is itself subject to a careful analysis in the opening chapters of *APo. Epistasthai* simpliciter is when we think we *gignōskein* the cause from which the thing results, that it is the cause and the fact cannot be otherwise. I agree with what I take to be the currently preferred view that understanding is a better rendering, in English, of Aristotelian *epistasthai* in *APo.* But we should be a little careful. 71b16f. raises the question of whether there is another manner of *epistasthai* and he owes us, and eventually gives, an account of the mode of cognition of the *archai* of demonstrations. The preferred term for that in II 19 is *nous* and the final paragraph of the treatise positively *contrasts nous* (of the principles) with *epistēmē* in the statement (admittedly in the conditional, *eiē* 100b11) that there is no *epistēmē* of the *archai*. However in the initial discussion of the topic in *APo.* 1 2 and 3 he not only asks whether there is another manner of *epistasthai* but at 72b18–20 uses an expression that seems to imply an *epistēmē* (no doubt in a loose sense) of the immediate premises that is not by demonstration.[8] However although there may be some slight looseness in the terminology, the *theory* is again in no doubt at all. Demonstration, we can be confident, is syllogistic and yields the kind of understanding that Aristotle specifies. It depends on primary premises that are themselves indemonstrable, the object of a different mode of cognition, *nous*.

---

[8] See *APo.* 72b18–20: ἡμεῖς δέ φαμεν οὔτε πᾶσαν ἐπιστήμην ἀποδεικτικὴν εἶναι, ἀλλὰ τὴν τῶν ἀμέσων ἀναπόδεικτον.

Those primary premises have to meet the conditions set out in I 2–4, and familiar as they are, they still take one's breath away. 71b20ff. states that they must be true, primary, immediate, more familiar than, prior to, and causes of the conclusion, and goes on to gloss more familiar than as what is so *phusei* rather than merely to us. I 4 further specifies that there is demonstration only of the necessary and lays down the requirements that the connections should be not just *per se* and universal but, as used to be said, 'commensurately universal'. We have three types of primary indemonstrables identified in I 2, axioms, definitions, and hypotheses/suppositions, and the relation between definitions and demonstration occupies him for much of book II.

This, the official theory of demonstration (as I shall call it) is nothing if not rigorous, and it is no wonder that there are such problems with regard to its applicability. Quite apart from the requirement that the reasoning be syllogistic, the conditions specified for the ultimate primary premises are very stringent indeed. How far the physical treatises, or any other works, operate with that theory of demonstration – with those requirements for the mode of deductive reasoning, with those conditions for the ultimate primary premises – how far those other works even presuppose that theory as an ideal, are questions we shall eventually come to. But first we should consider whether there is not evidence, within *APo.* itself, of some relaxation of those ultra-strict conditions set out in the opening chapters. How far are there signs that in the *APo.* itself Aristotle is prepared to broaden the scope of *apodeixis* with a corresponding relaxation of its conditions?

The answer is that there are a few such signs, but one has to say they do not amount to a great deal. The most important concession is the well-known distinction between the *apodeixis* of the *hoti* and the *apodeixis* of the *dioti* presented in *APo.* I 13, I 27 and II 16 especially. Whereas initially *apodeiktikē epistēmē* proceeds from premises that are causes of their conclusions, 71b22, it turns out that you can have not just *sullogismos*, but an *apodeixis* (of a kind) when you do *not* take a middle term that specifies the cause. You can prove that planets do not twinkle from their being near (where this is through the cause): but you can also prove, have an *apodeixis* of, their being near from their not twinkling (when that is not through the cause) (*apodedeiktai* at 78a36, *apodeixis* at b14). A second concession or relaxation is implied by the comparison between more, and less, perfect modes of demonstration in I 24–6.* *Reductio ad impossibile* is inferior to ostensive demonstration but still called *hē eis to adunaton apodeixis* at 87a15f., and negative demonstrations are inferior to affirmative ones, just as particular (that is, less universal) ones to (more) universal demonstrations (I 24). But I think it is important to see that though there are, as he puts it at 86a17, more exact *akribestera* and (by implication) less exact demonstrations, for example depending on how close to the *archē* they are (cf. 74a10), this implies no strategic modification to the model of demonstration itself.

Of course in *APo.* II especially examples of deductive inference that do *not* meet the requirements of the official theory of demonstration do occur with some frequency. There are notorious problems with syllogisms that have premises that are true only for the most part (I 30, II 12), though they are still *syllogistic* in form.⁹ Worse still there are examples of syllogisms where one or more term is particular (the Persian invasion II 11, the Nile II 15) though the problems these may be thought to present are not taken up and do not lead Aristotle to modify his syllogistic. Moreover whatever relaxations to the model of reasoning they might have suggested, they carry no implications for the theory of demonstration, since *apodeixis* is not the term used.

But what about the exploration of a variety of astronomical and botanical problems in *APo.* II 13–18? Do not these show that Aristotle sought at that time to apply his theory of demonstration to physics? Indeed they may: but we may feel that any actual application still presents greater difficulties than are recognised in the text. Even if it is explained to us that deciduousness may be defined in terms of the coagulation of the sap at the junction of the leaf and stalk, there are still plenty of points of unclarity about the terms Aristotle uses. How is 'broad-leaved' to be defined? How broad is broad? Again what about the 'nearness' of the planets? How near is near? 'Non-twinkling' is not exactly satisfactory either – even if we allow (on the pragmatic principle of relevance,

---

⁹ See especially Mignucci 1981.**

* *APr.* I 23, 40b23ff., 41b1ff., had already referred to a variety of modes of *apodeixis* and *sullogismos*, including negative, particular and hypothetical ones, in its proof that *all* syllogisms can be reduced to the three figures Aristotle has identified.

** Cf. below, ch. 7 at n. 5.

helped by the distinction between *mē* and *ou*)[10] that it is not *anything* of which 'it does not twinkle' is true that is in mind, but only certain subjects, namely sources of light. Even with that specification, neither universal proposition (what does not twinkle is near: what is near does not twinkle) is unexceptionable and for reasons that go beyond the vagueness of 'near'.

So these syllogisms will certainly need a lot of tidying up to be at all satisfactory and of course the standard response to the objection is to say that they are only intended to illustrate the *types of relationship* Aristotle is interested in, not to be themselves valid inferences or actual demonstrations. However the problem does not go away. In the course of analysing, from II 13 onwards, how we 'hunt for the attributes in the *ti esti*', Aristotle uses a number of examples that *might* have led him to reflect on the difficulties of applying his ideal model of demonstration in practice to physics: and *we* may well wonder just how satisfied Aristotle himself was with the sample explanations offered. But as for his actually being led by any such reflection to modify or qualify the ideal model, in particular by relaxing some of the more stringent conditions on the primary premises of demonstrations, there are – we must say – no signs of his doing so *here*. Whatever *we* may think of the evident strains present in any application of the model to physical examples, there is no indication that *Aristotle's own* reaction – in the *APo.* – to whatever strains he perceived was to modify the model itself substantially in the later chapters of that work.

But how does the official theory of demonstration fare elsewhere in Aristotle? Quite often in the scholarly literature the analysis proceeds by comparing *APo.* with the physical treatises directly. Yet (as I have indicated) that is to jump the gun; for there is much that is germane to his views both on the theory and on the practice of *apodeixis* in other works as well. Questions can already be raised with regard to the *Topics*. As is well known, the opening chapter sets out a looser or more general account of *sullogismos* than that used in the *Analytics*. In the *Topics* (100a25ff.) *sullogismos* is a *logos* where when certain things are laid down, others follow through what is laid down – a definition evidently compatible with other styles of deductive reasoning besides Aristotelian syllogistic – and a text that has, accordingly, been given pride of place in arguments to the conclusion that there is a pre-syllogistic use of *sullo-*

[10] E.g. at *APo.* 78a34.

gismos present, in particular, in such passages as this. Hard on the heels of that statement we are told at *Top*. 100a27ff. that there is *apodeixis* when there is *sullogismos* from true and primary premises or from premises that themselves are secured from true and primary ones.

To be sure, Aristotle does not have to say everything that may be relevant to *apodeixis* every time he explains the term, and on that basis we might want to insist that the *Topics* statement is perfectly compatible with the definition of *apodeixis* in *APo*. However it is also obvious that if compatible, it is still *far less complete*, especially in that it makes no mention of the other conditions the *APo*. lays down for the primary premises, such as their being more familiar *phusei* than, prior to, and explanatory of the conclusions (cf. however *Top*. 141a29ff.). *Topics* I 1 describes the *prōta* as commanding conviction through themselves, and again that is far less explicit than *APo*.: and it is perhaps particularly striking that where the *Topics* appears to be satisfied with mentioning the two requirements, true and primary, a text in *APo*. says in so many words that that is not enough, adding that there must be *suggeneia* (*APo*. I 9, 76a28–30). So those who would see a pre-syllogistic notion of *apodeixis* as well as of *sullogismos* in play in the *Topics* may emphasise the lack of explicit mention of key characteristics of the *APo*. model, even though unitarians may counter that while less explicit than *APo*. *Topics* I 1 is, as far as it goes, compatible with it. Certainly on any account that chapter operates with a sharp contrast between demonstration and dialectical syllogisms, based on *endoxa*, 100a29ff.

Yet elsewhere in the treatise we have to acknowledge that what is called *apodeixis* sometimes does *not* meet the conditions laid down in *APo*. Let me pass briefly in review three of the more remarkable such texts. First 105a7ff. offers advice about the problems the dialectician should discuss or rather more specifically about those he has no need to consider. We ought not, he says, to discuss subjects the *apodeixis* of which is too close or too distant, not too close because they raise no *aporia* whereas the distant ones involve more than is a matter of training, *gumnastikē*. But that may suggest a rather different relationship between dialectic and demonstration than is suggested by the distinction in their premises: it is as if between the two areas, of the non-aporetic and the ultra-aporetic, there is third domain – a possible domain of *apodeixis* it would seem – which the dialectician is positively encouraged to tackle.

Here then may be a first softening of the exclusive contrast between dialectic and demonstration.

Then secondly, still in the first book, we have a reference to syllogisms, and indeed to an *apodeixis* described as *ex hupotheseōs* (108b7ff., 19). You first gain admission that what holds of certain similar cases, holds also of the case in question. Then proving the former, you thereby prove the latter: indeed you have carried out the *apodeixis* on the hypothesis set out in the preliminary admission. Yet this is no *apodeixis* at all, by the standards of the official theory: for this hypothesis is no indemonstrable primary premise, but merely an admission you obtain (and cf. the criticisms of hypothetical *apodeixeis* in relation to essence in *APo.* II 6, 92a6ff.).

The third text is even more striking. At *Topics* II 4 we have a *topos* designed to show the presence of contraries. Examine the genus of the species in question and then move to the species itself (111a14ff.): in the example, to show that rightness and wrongness belong to perception, consider judgement,* for that is the genus of which perception is a species. That is said to be an *apodeixis* from the genus to the species (111a18f.): yet the sequel admits that this does not hold generally in constructive arguments (though his example was a constructive argument). It is particularly striking that the move here described as *apodeixis* is (in the very form in which it is exemplified) not valid. Thus although *Topics* I 1 might seem to seek to reserve *apodeixis* for a domain that is itself strongly contrasted with dialectic, other later uses of the term are a good deal looser, and indeed in the *de Sophisticis Elenchis* (167b8ff.) there is already explicit recognition that there are *rhetorical apodeixeis* when Aristotle mentions those from signs in his discussion of the fallacy of the consequent.

It is, of course, the *Rhetoric* itself that presents us with a second fully elaborated theory of – precisely – rhetorical *apodeixis*. Clearly if we are to do justice to Aristotle's theories and practices of *apodeixis* this evidence cannot be dismissed, any more than the materials in the *Topics* can: and indeed there are points of some relevance to our interpretation of particular passages in the physical treatises that emerge from the *Rhetoric*. As is well known, the *Rhetoric* presents us with a theory of the enthymeme as rhetorical demonstration. *Rh.* 1355a4ff. first says that *pistis* is a demonstration – only *apodeixis tis* of course, but a demonstration of a kind never-

---

* Or rather discrimination, cf. below ch. 6.

theless – for we believe especially when we suppose something to have been demonstrated *apodedeichthai* – and rhetorical demonstration is the enthymeme, which is the most proper or most powerful, *kuriōtaton*, of *pisteis*. Much of the subsequent analysis of argument in the *Rhetoric* centres either on the enthymeme as the counterpart, in rhetoric, of *sullogismos* or on *paradeigma*, the counterpart of *epagōgē*. There are, to be sure, complexities and puzzles in his theory of enthymeme, studied recently by Burnyeat,[11] about how it came to be called such, and in particular on the notion of *sullogismos* in play when enthymeme is said to be its counterpart: have we the fully fledged syllogistic in the background or just deductive reasoning?

However those and other questions do not affect the substantial point we are concerned with here – which is that Aristotle has no hesitation in applying the terms *apodeixis, apodeiknumi*, and so on, within rhetoric, marked off as *being* rhetorical, but demonstration nonetheless. Now among the points of general similarity between rhetorical demonstration and the notion of demonstration analysed in the *Analytics* we may count first that it secures conviction. As 1355a4ff. puts it, it is the chief way of doing so, though there are others as well that do not depend on argument at all, but rather, for example, on the character of the speaker or on those of the hearers (cf. e.g. II 1, 1378a6ff., III 1, 1403b11ff.). Moreover *Rh.* II 25, 1403a10ff., using the *Analytics*, points out that *tekmēria* and enthymemes using them cannot be refuted on the basis of not being consequential. All you can do to object to them is to show that the alleged fact is no fact: but if there is no doubt about it (the fact), then refutation is impossible.

However so far as major divergences go, demonstration in the *Rhetoric* relates sometimes, and maybe more particularly, to what is *unclear*. You can attempt demonstration in relation to the past, for example – indeed 1368a32f. says that the past admits *apodeixis* especially *since* it is unclear – and again you will need demonstration with regard to paradoxes (1394b8ff). In general in *Rh.* III 17 he remarks first that *pisteis* should be demonstrative (*apodeiktikai*, 1417b21), then that it is necessary to demonstrate (*apodeiknunai*) when there is dispute about four types of matters, first as to the fact, second as to the question of whether harm was done, third as to the degree of harm, fourth as to the justification for it; and he

---

[11] See Burnyeat 1994.

goes on to remark that in epideictic speeches when the facts *are* clear, there is seldom need to demonstrate them: 1417b32ff.

There is much that is quite sensible, much that is quite wily, in the discussion of demonstration in the *Rhetoric*, but of course not only are the contexts of exchange (deliberative, forensic, epideictic oratory) quite different from those presupposed in the *Analytics*, but one might almost say that the occasions for demonstration are the very reverse of those envisaged in the *Analytics*. If there were an occasion when a forensic *orator* had premises that were true, primary, immediate, better known than, prior to, and explanatory of the conclusions, the last thing he would dream of needing to do is to proceed to a demonstration. It is precisely when the facts or their interpretation are highly disputed that demonstrative arguments – rhetorical style – are needed, and though they may be hard to give, there were other means of conviction to hand to support or replace them – your character and your effect on your audience, your aim being, of course, just to get *them* to believe.

So far we have considered the divergent theories and uses of *apodeixis* in the *Organon* and the *Rhetoric* – and the differences are certainly as remarkable as the similarities. But it is still not yet time for us to turn to the physical treatises, for both the ethical works and the *Metaphysics* provide further considerable material for our study. We must be even more selective than before and I shall try to summarise the main points as concisely as possible.

Several very well-known texts in the *Nicomachean Ethics* take their stand by the *Analytics* notion of *apodeixis*, to the point that one might say that if one just had the *EN* and the *Analytics* to go by, one could be satisfied that Aristotle never strayed from the official theory. The discussion of the intellectual virtues in *EN* vi (= *EE* v) cites the *Analytics* and repeats its distinction between cognition of the *archai* (assigned at first to *epagōgē*, 1139b26ff., then associated with the faculty of *nous*, 1140b31–1141a8, 1143a35ff., b10ff.) and the *epistēmē* itself, termed a *hexis apodeiktikē* at 1139b31f. This is an analysis where he is being precise and not just following similarities (1139b18ff.). When he comes, in the next book, to give his famous methodological recommendation, 1145b2ff., that with certain subject-matter a point has been sufficiently shown if the difficulties have been resolved and the *endoxa* are left standing, the verb he uses (that I have translated 'shown') is not *apodeiknumi* but *deiknumi* (*dedeigmenon*, b7). So far is *EN* i from the *Rhetoric* theory of demonstration that at 1094b25ff., in an equally famous passage in which he illustrates the different degrees of exactness to be expected in different inquiries, Aristotle says it would be similar (by which we understand similarly absurd) to allow a mathematician to speak merely plausibly as to demand *apodeixis* of a rhetorician. True enough, of course, if we take demonstration in the *Analytics* sense: yet in the *Rhetoric*, as we have seen, Aristotle had plenty to say about the kind of *apodeixeis* that the rhetorician can and does deploy.

One text in the first book of the *Eudemian Ethics*, however, may represent a departure from the *Analytics*. *EE* I 3, 1214b27ff., says that it is superfluous to examine *all* opinions about happiness, that is including, for instance, those of children, the sick, and the mad. It is sensible to examine the most notable alone (cf. also *EN* 1098b11ff., 27ff.). There are difficulties that are proper to each inquiry and so also concerning the best life. But it is a good idea to examine those opinions, since 'the refutations of opposed arguments are demonstrations, *apodeixeis*, viz. of the argument itself' (1215a7ff.). Now it is true, of course, of strict contradictories that the disproof of one *is* the proof of its contradictory – the principle on which *reductio* depends. But first we have seen that *APo.* rates *reductio* below direct demonstration (I 26, 87a1ff.). Second and more importantly, it is not as if the opinions to be examined on the best life were formulated in crisp terms such that the refutation of p is enough to establish not p. Those opinions took the form of beliefs such as that prudence is the greatest good, or virtue is, or that prudence is a greater, or lesser, good than virtue and so on (1214a32ff.). But that is rather unlike the case mentioned, for instance at *SE* 170a23ff., namely the opinion that the side and the diagonal of the square are commensurable, the refutation of which does indeed show that they are incommensurable. Third, even if such a refutation of an ethical belief established the *truth* of the contradictory, there are still other requirements on demonstration set out in the *Analytics* that should, but are unlikely to, be met, namely that the conclusion should be shown via premises that are primary and explanatory of it. It is, then, rather in a more general – if not necessarily precisely the *Rhetorical* – sense that we should take the reference to *apodeixeis* in *EE* I 3.

That reading gains support from some prominent texts in the *Metaphysics*. Plenty of passages, there too, are compatible with the most rigorous notion of demonstration, as set out in the *Analytics*, especially a sequence of texts that show that there can be no *apodeixis* of the essence (*ti esti*, e.g. *Metaph.* 997a31), nor of particular perceptible substances (1039b27–1040a5) and that recognise in general that the principles of demonstrations are not demonstrations (1011a13, cf. 1013a15). On the other side we also find Aristotle perfectly prepared to use the term *apodeixis* of his predecessors' or opponents' forays into the field, of their *attempted* demonstrations of their own metaphysical positions (e.g. 990a24, 1000a19f., 1087b21). It is not that those texts should be taken as reports of *their* terminology, that is as evidence of pre-Aristotelian usages of *apodeixis*. But no more is it the case that their arguments conform to Aristotle's own *APo.* requirements. Yet he thinks it appropriate to evaluate them as – purported – *apodeixeis* nevertheless, and for that to *be* appropriate, it had better be the case that those arguments had some pretensions to be demonstrations in *some* sense. Evidently the less stringent the conditions imposed on what is to count as (in some sense) a demonstration, the more plausible, the more justifiable, it is for Aristotle to use that terminology in referring to them.

Yet it is rather in two further connections that the *Metaphysics* provides particularly important evidence for our analysis, first the well-known discussion of how to recommend the general axioms of reasoning, the laws of non-contradiction and of excluded middle, and secondly in the brief account of the variety of modes of demonstration that we find in E 1. In Γ 4 Aristotle's opening gambit is to say that it is a mark of a lack of education not to know what things need demonstrations and what do not (1006a5ff.). As regards the law of non-contradiction, he asks, what principle could it be proved from (1006a10f.)? However an elenchtic demonstration is possible if the person who challenges the principle just signifies something (*sēmainein ti*, a21). In the exploration of the distinction between elenchtic demonstration and (just) demonstration, he points out that the latter is not possible here without begging the question (for example if you demanded that your opponent *asserted* something). But provided the discussion is started by someone else, there can be refutation. At 1006a17f. he says that this is refutation (*elenchos*) *and not apodeixis* and that is what we should expect in the light of the official theory of *APo*. But he had earlier said that the principle can be given an elenchtic *demonstration* (1006a11f.), implying a wider category of qualified demonstration, and that is not superseded by 1006a17f. since at a24 he repeats that there will be a demonstration (of the principle) to wit in that elenchtic sense, once the opponent has admitted something definite. Similarly when these problems are discussed again in K (whose authenticity is, to be sure, not uncontroversial) we have a similar distinction stated in slightly different terms, when he remarks of the fundamental principles that there is no unqualified (*haplōs*) *apodeixis*, but there is one *ad hominem* (*pros tonde*, 1062a3ff., a5ff., a9ff. – alone – a30ff.).

But if Γ and K allow qualified demonstrations as well as unqualified ones (cf. also Δ 1015b6ff.), the door is opened wider still and in a more general context in E 1, perhaps the most important text for the purpose of our understanding of demonstrations in physics. E 1 offers a general categorization of the various *epistēmai*, first noting differences in the first principles and causes they deal with – for these may be more or less exact (*akribesterai*) and more or less unqualified (*haplousterai*). But none of them undertakes a *logos* of the essence, and all assume some genus of being. Starting from this (the essence) some make it clear to perception, others take the *ti esti* as a hypothesis, but what both types do[12] is to demonstrate the essential attributes of the genus they deal with – the *kath' hauta huparchonta* – though they may do so in a more necessary, or in a laxer, manner, *anankaioteron, malakōteron*, 1025b13. That he has *phusikē* in mind as one such *epistēmē* (whatever its precise categorisation should be) is clear from the immediately following passage, which refers to it as dealing with the *ousia* that is such as to have an *archē kinēseōs kai staseōs* in itself, 1025b18ff.

Where this text corresponds precisely and where only more loosely to *APo*. are alike significant. Of course it is standard official theory that the *archai* must themselves be in some sense assumed, and there must be a genus about which the demonstrations will be the demonstrations they are. Those sciences that take the *ti esti* as a hypothesis fit the schema of *APo*. 1 2 well enough, and although there may be some puzzlement over the alternative expressed here, when it is said that some sciences make it – presumably the *ti esti* – clear *to* perception, *aisthēsei*, 1025b10f., we can perhaps relate this to those texts such as *APo*. 78a34f. that speak of grasping certain premises *by means of* perception (which is there mentioned as an

---

[12] *hai men* as well as *hai de* are subjects of *apodeiknuousi* at *Metaph*. 1025b10ff.

alternative to induction, or perhaps a specification of it, for perception does of course figure prominently in his account of induction in *APo.* II 19).

But more important, so far as the demonstrations themselves go, these are said in E 1 to be more or less necessary, more or less lax. Of course more or less exactness in the principles of the sciences (the point at 1025b7) is readily intelligible in the light of the elucidation of that in terms of the degree of abstractness and of priority in definition in such texts as *Metaph.* 1078a9ff. and cf. 982a25ff. From the *APo.* itself we can cite the discussion of the degree of exactness of the sciences at I 27, 87a31ff., and in the passage I mentioned before, 86a16ff, *APo.* speaks of more and less exact demonstrations, depending on how close to the *archē* they are. But *Metaph.* 1025b13 goes beyond *APo.* in allowing greater or less necessity, more or less laxity, in the demonstrations. This is not the same point as is made in *APo.* I 24–6 when those chapters distinguish between superior and inferior *apodeixis* according to whether they are direct or use *reductio*, affirmative or negative, universal or particular (in the sense of less universal), for none of those goes so far as to imply that the inferior demonstrations need not meet the condition set out in *APo.* I 4 that demonstration is confined to what is necessary. Yet in E 1 in the *Metaphysics* we have demonstrations that are allowed to be less than totally rigorous.* It is not as if, in the looser demonstrations, there is some – loose – demonstration of the *essence* – for that is ruled out at 1025b14ff. But we are here allowed more or less rigorous demonstrations of the essential attributes, depending on the science, and the importance of that will become clear as we proceed.

So at last – and not before time – we may turn to the physical

---

* The reasons for the laxity in the demonstrations are not detailed in *Metaphysics* E 1, so supplying them is hazardous and conjectural. But we should bear in mind that Aristotle is here concerned, among other things, to argue for the superiority of the supreme theoretical inquiry, here called 'theology'. Physics, in this context, is said to deal with items that, like 'snub', are taken with the matter (cf. below ch. 2 at n. 49). That gives an element of indeterminacy in its subject-matter, as also does the fact that physics deals with what is true not just 'always' but 'for the most part' (mentioned in E 2 at 1026b27ff., 1027a8ff.), though as already noted (n. 9 above) the problem of taking 'for the most part' in a statistical sense is that syllogisms with both premises true 'for the most part' will not yield conclusions true 'for the most part', let alone 'universally' (see further below ch. 7 at n. 5). Aristotle's problematic, in *Metaphysics* E 1, thus seems to differ from that in *PA* I 1, to be considered below, even though both texts countenance a pluralism within what count as modes of demonstration.

treatises themselves, though I hope now with a firmer idea of the complex possible range of uses of our key terms *apodeixis* and *apodeiknumi*. In the physical works those terms figure both in a number of important methodological passages and in other contexts where Aristotle is concerned not so much with methods as with results, and the first point that is worth remarking is that though we have a fair number (some thirty) of texts to consider, they are nothing like so common as one might have expected from the importance and frequency of the terms in the *Analytics*. The vocabulary of causation, by contrast, and references to the natures of things and to their *archai* are very heavily used in the physical works from the *Physics* through to *GA*.

Moreover a number of texts in which *apodeixis* or *apodeiknumi* figure do not contribute a great deal to our understanding of the extent to which, in the physical works, Aristotle operates with what I called the official theory of the *APo*. That is true, for instance, of some of the passages where the 'demonstrations' in question are not his own, but those that other theorists had put forward (not necessarily *calling* them such, of course) generally considered by Aristotle himself to be botched jobs at best. One such text is *Cael.* 279b33ff., 280a5, where he refers to the defence of Plato's position that the world is generated but indestructible that has it that that was just for the sake of instruction, like geometrical diagrams, where Aristotle goes on to comment that in their *apodeixeis* the situation is quite different – for unlike what is true in geometrical diagrams, in *their* case in the attempted defence of this cosmological position, what is laid down and what is concluded are quite different, indeed the antecedents and the consequents contradict one another. Obviously Aristotle does not endorse the Platonic defence, but for that very reason it is difficult to say whether he is evaluating their position by the strictest standards of *APo*. demonstration (when their arguments are not demonstrations at all) or whether he has merely a looser sense ('exposition') in mind.

There are indeterminacies, too, in some of the passages where the terms are used of sequences of arguments that Aristotle does offer in his own person. A number of texts in the *Physics*, for instance, express little more than a general concern for the deductive structure of his arguments. Once he has established, to his own satisfaction at least, certain preliminary points, he moves on to further points to be considered on the basis of what has already been demonstrated. Several such examples appear in his discussion of *kinēsis*, and especially of the problem of infinite movement, in *Physics* VI, e.g. *Phys.* 233a7, 238a32, b26, 240b8, cf. also 233b14, 237a35, 238b16, passages that illustrate, among other things, how what is claimed to have been demonstrated has not always been shown systematically by syllogistic arguments starting from explicit primary premises.

Elsewhere in the physical treatises there are texts that are certainly compatible with the *APo.* model of demonstration and many may think that the most economical hypothesis is to assume that Aristotle has that model in mind. Thus *de An.* 402b25 states that the *archē* of every demonstration is the essence, and 407a25f, developing a criticism of Plato on the soul, first states – a touch extravagantly – that all *logos* is either definition or demonstration, and then specifies, quite understandably, that *apodeixis* has both a beginning and an end (and so is not the kind of eternal revolving movement that Aristotle is concerned to refute).

Yet we had better be careful about inferring too swiftly that the official theory is in the background in such a text as *HA* 491a11ff., a passage of which much has been made by some who have argued for that interpretation. There in a methodologically self-conscious text Aristotle specifies that we must first grasp the existing differences and *ta sumbebēkota pasi*, and then try to find (*heurein*) their causes. That is the natural method of procedure, once a *historia* concerning each thing is to hand. For that concerning which (*peri hōn*) and from which (*ex hōn*) one should make the *apodeixis* become clear from these (that is, I take it, from the results of the *historia* setting out the differences and the *sumbebēkota*). All of that may look rather grist to the mill of those who want to see *APo.* style demonstration as the aim of the zoology. Yet caution dictates that we should remark that the terminology here used of the subject (the *peri hōn*) and the starting-points (*ex hōn*) of demonstrations is *neutral* as between different styles of demonstration. Even the *Rhetoric*, after all, speaks of the *peri hou* and indeed of the *archē* of rhetorical demonstrations (1414a30, 1418a26): *any* demonstration, clearly, has to have a subject-matter and starting-points, and reference merely to that fact does not, of itself, tell us *what* style of demonstration is in question.

Critics may at this point object that that is to be excessively cautious, and that when we find, in the physical treatises, ideas and terminology that *fit* the *APo.* model of *apodeixis* we can presume that Aristotle wants to apply it and is doing so. However that we cannot adopt that presumption as a general principle is I think abundantly clear. Let me take five passages in particular which serve to illustrate the point.

First there is a striking text in the *Meteorologica* I 7, in the account of comets. Having run through a number of previous theories on the subject in chapter 6, he begins chapter 7 with the following statement, 344a5ff.: 'Since concerning things that are unclear (*aphanē*) to perception we think that the matter has been sufficiently *demonstrated* according to *logos* if we bring it back to what is possible, and proceeding from the present *phainomena* perhaps one might think that it happens like this ...' This is similar to a number of other passages,[13] in which Aristotle remarks, in relation to certain difficult or problematic subjects, that a satisfactory outcome to an inquiry will settle for *less* than total elucidation of the problem. Yet where some such texts use the term *deiknumi* in one or other of its forms, the *Meteor.* passage is exceptional in talking of a sufficient demonstration: *hikanōs apodedeichthai*, 344a5f. Yet a demonstration that merely states what is possible is, of course no *APo.* style *apodeixis*. Aristotle does not here specify exactly what will count as 'what is unclear to perception', and if comets are, maybe a lot more also is. But in such cases, evidently, whatever they are, the principle here enunciated states *not* (as we might expect from *APo.*) that on certain obscure topics there are no demonstrations at all, but that what counts as a sufficient demonstration, in their case, is to have given a *possible* account.*

My second passage is *Cael.* 279b4ff. which deals with the generability/destructibility of the cosmos. We should first go through the opinions of others, he says, for 'the demonstrations *tōn enantiōn* are difficulties concerning *enantiōn*'. Guthrie takes the first *enantiōn* as well as the second, to refer to contrary theses or views and trans-

---

[13] One such text, *EN* 1145b2ff., has been noted. Cf. also *Phys.* 211a7ff.

* The sense of 'sufficient' seems then to be 'as good as can be expected, given the difficulties of the inquiry'. It is not that this is sufficient by the criteria of demonstrative understanding set up in the opening chapters of *APo.*, for then there would be no need to apologise for it. But it is not just sufficient *ad hominem*, since the context is not the persuasion of a particular audience.

lates: 'since to expound one theory is to raise the difficulties involved in its contrary'.[14] It is possible however that the first should be taken as masculine: the demonstrations of opponents are the sources of difficulties concerning contraries. This might bring it close to the passage from *EE* we considered (1215a7ff.) where the context is broadly similar. Be that as it may, *Cael.* 279b7ff. proceeds (in Guthrie's translation): 'At the same time also the arguments which are to follow will inspire more confidence if the pleas of those who dispute them have been heard first. It will not look so much as if we are procuring judgement by default. And indeed it is arbiters (*diaitētai*) not litigants (*antidikoi*) who are wanted for the obtaining of an adequate recognition of the truth.' True, this introduces a second point (*hama de*). But it strongly suggests that whatever the *ideal* for argument in the physical treatises may be, in *practice* some of the discussion will be (indeed it is) at the level of dialectical debate with opponents – even if as 'arbiters', not as 'litigants'. Moreover whichever way we take the first *enantiōn* at 279b4ff., whether as masculine or as neuter, the *apodeixeis* in question there are evidently quite unlike the demonstrations we meet in *APo.* in one obvious and undisputed respect, namely that these are demonstrations that are themselves *aporiai*. Where *EE* spoke of the demonstration of one position being obtained via the refutation of its opposite, *Cael.* has demonstrations being the source of difficulties for those who take the opposite view.

My third text, *GA* 747b27ff., like the first, comes in a passage where he has just been criticising the theories of his predecessors. *GA* 747a28 criticises Empedocles and Democritus for attempting too global an *apodeixis* of the sterility of mules, one that covers all cases of copulation between animals of different kinds. Aristotle's own first attempt at a resolution is introduced at b27ff.: 'perhaps a *logikē apodeixis* would seem to be more trustworthy than those we have mentioned', and he glosses *logikē* immediately as one that is more universal and further from the proper or appropriate *archai*. There then follows a dilemmatic argument to the effect that mules cannot produce *either* offspring of a *different* kind (because the offspring of a male and a female of the same species belongs to that species) *or* of their own kind (because a mule is the offspring –

---

[14] See Guthrie 1939. The 'demonstrations' in question are, presumably, those connected with what he has just referred to as the 'suppositions', *hupolēpseis*, of other theorists.

not of two mules – but of a horse and an ass). That leads to the comment, 748a7ff., that this argument is excessively general and empty: arguments that do not start from appropriate principles may appear to be to the point, but are not. Worse, this one is not even true (748a12) – for many animals produced by parents of different species are fertile. We must, to be sure, pay due attention to the qualification *logikē* when the argument is introduced as a *logikē apodeixis*. There is no question of this being a demonstration in an unqualified sense – and no question, come to that, of Aristotle endorsing the argument it contains. But it is still striking enough that he does not just deploy his usual distinction between investigations that proceed *phusikōs* and those that proceed *logikōs*. Here he is prepared to combine *logikē* (in a pejorative or at least non-approved sense) with *apodeixis* and this suggests that the latter term covers a spectrum that includes the *APo.* model at one end, but extends as far as merely apparent demonstrations at the other.

GA 742b17–35, my fourth text, also contains surprises and difficulties for any who seek too close a fit between the practice of *GA* and the requirements of *APo.* Democritus is criticised for rejecting the investigation of the *dia ti* beyond the point where he had stated that the explanation consists in something that 'happens always'. What is limitless has no starting-point. But the cause is a starting-point and what is always is limitless. So, according to Democritus, to ask for a cause in connection with what always is is to try to discover the *archē* of the *apeiron*. Now the consequence of that, Aristotle says, would be there would be no proof of things that are eternal: whereas there are plenty, both of things that come to be always and of things that are, e.g. geometrical proofs. For good *APo.*-style reasons, Aristotle concedes that there is no *apodeixis* in some cases, namely where you are dealing with the *archai* of demonstrations – of which there is, he says at 742b32f., another *gnōsis* and no *apodeixis*. So far so good and we understand the point to be that of the indemonstrable primary premises of demonstrations there are themselves no demonstrations: rather they *are* the *archai*. But then he continues, b33ff., that the *archē* in the case of what is immutable is the *ti esti*, while in the case of things that come to be there are several and their manner is different and they are not all of the same kind. Among these *archai* one is that from which movement begins and so all the blooded animals have a part that comes to be first, namely the heart.

Here he first distinguishes things of which demonstration is possible from the things (indemonstrable *archai*) on which those demonstrations depend, and while that gives him a point of agreement with Democritus on the principle that not everything can be demonstrated, he insists against Democritus that one can give demonstrations of such eternal geometrical truths as that the angles of a triangle sum to two rights or that the side and diagonal of the square are incommensurable. But by the end of the section the *archē* he is talking about is very different, no longer the indemonstrable starting points of demonstrations but the efficient cause: he has shifted from primary premises to the heart, the investigation of which he then duly pursues, for all the world as if he were investigating something analogous to (though different from) the *ti esti*. Now it is true, of course, that Aristotle explicitly distinguishes the *archē* in question in the case of what is immutable from that or those to be found in what comes to be, and that of course is in line with many other passages in Aristotle that alert us to the fact that *archē* is said in many ways (e.g. *Metaph.* 1012b34ff. especially). Yet the remarkable point here is that while distinct, there has to be some comparability (other than the mere name) between the two types of case for Aristotle's comment on Democritus to make sense. For the non-demonstrability of the indemonstrables is meant to be the sense in which Democritus is right, while the investigability of the efficient cause is one way in which he is wrong. While Aristotle *states* that, of things that come to be, their *archai* are several and their manner not all of the same kind, this is left *unexplained*: and he certainly makes no comment here whatsoever on the question of whether – with those things that come to be – there is demonstration of the type he has exemplified from geometry. That type of demonstration is of what is eternal, and that spans *both* what is and what comes to be *always* – the eternal heavenly bodies, we may presume. But so far as this text is concerned, we are given no inkling of the possibility of geometrical-style demonstrations with regard to sublunary phenomena.

My fifth and final text is at once the most difficult, the most controversial and the most enlightening. This is the famous or notorious discussion of the very difficulty I have just mentioned, the manner of demonstration appropriate to and possible in physics that comes in the opening chapter of *PA* 1. This deserves a separate paper to itself (indeed Düring devoted one to it in the 1960

Symposium Aristotelicum – Düring 1961), but I can omit at least some of the detailed disputes since for my present purposes it will suffice to point out that *whatever* view is taken on the main controversial issues there are important conclusions to be drawn concerning Aristotle's theories – and his practices – of demonstration.

First however I have to outline the overall structure and context of Aristotle's argument, and recall where the chief disputes in interpretation arise. *PA* 1 1 is of course *devoted to setting out the right method*, a matter on which Aristotle complains (639b5ff.) there had previously been a good deal of indeterminacy. 639b11ff. introduces the point that there are several types of cause, e.g. final and efficient, and illustrates this from the works of art as well as those of nature. 639b21ff. further distinguishes types of necessity, the unconditional present in what is eternal, and the conditional necessity in things that come to be, including again artefacts. Taking the example of a house Aristotle says that for the end to be realised, there must be matter of a certain kind and certain antecedent processes – first one thing comes to be and is moved, then another. Similarly, 639b30, with the things that come to be *phusei*. 'But', he continues, 'the manner of the demonstration and the necessity is other in the case of physics and the theoretical sciences [my translation reflects the ambiguities of the Greek]. This has been discussed elsewhere. But the *archē* in some cases is what is, in others what will be. For since health or man is such, necessarily this [so-and-so] is or comes to be: it is not that because this [so-and-so] is or came to be, necessarily it [the product] is or will be. Nor is it possible to join the necessity of such a demonstration (*tēs toiautēs apodeixeōs*) to the eternal, so as to say, since this is, that this is. I have discussed these matters in another place...'[15]

The major dispute in the interpretation arises over the contrast intended in the ambiguous sentence (639b30–640a2) I duly translated ambiguously. Is the contrast between 'physics' on the one hand and the 'theoretical sciences' on the other (I shall call this option 1).[16] Or is it one between those two taken together (on

---

[15] *PA* 639b30–640a8: ἀλλ' ὁ τρόπος τῆς ἀποδείξεως καὶ τῆς ἀνάγκης ἕτερος ἐπί τε τῆς φυσικῆς καὶ τῶν θεωρητικῶν ἐπιστημῶν. (εἴρηται δ' ἐν ἑτέροις περὶ τούτων.) ἡ γὰρ ἀρχὴ τοῖς μὲν τὸ ὄν, τοῖς δὲ τὸ ἐσόμενον. ἐπεὶ γὰρ τοιόνδ' ἐστὶν ἡ ὑγίεια ἢ ὁ ἄνθρωπος, ἀνάγκη τόδ' εἶναι ἢ γενέσθαι, ἀλλ' οὐκ ἐπεὶ τόδ' ἐστὶν ἢ γέγονεν, ἐκεῖνο ἐξ ἀνάγκης ἐστὶν ἢ ἔσται. οὐδ' ἔστιν εἰς ἀίδιον συναρτῆσαι τῆς τοιαύτης ἀποδείξεως τὴν ἀνάγκην, ὥστ' εἰπεῖν, ἐπεὶ τόδ' ἐστίν, ὅτι τόδ' ἐστίν. διώρισται δὲ καὶ περὶ τούτων ἐν ἑτέροις.

[16] See, for example, Balme 1972, *ad* 640a1.

the one hand) and the products of *technē* (on the other) – call that option 2a.[17] Or, a variation on that, option 2b,[18] a contrast between physics and the theoretical sciences on the one hand, and the products of *technē* and the things that come to be *phusei* on the other. In that last option, note that *phusikē* comes on one side but things that are or come to be *phusei* on the other.

Now each of these readings has its strong points, each has its weaknesses. Thus option 1 faces the embarrassment that usually physics is itself treated as a theoretical science, as most notably at *Metaphysics* 1025b24ff.: so to rescue that view one has to understand the contrast as being between physics and the *other* theoretical sciences. Option 2a, as I stated it, suffers from the problem that when Aristotle illustrates the type of necessity he is concerned with, in the case of things that come to be, his two examples are health and *man*, and man cannot by any stretch of the imagination be deemed a product of *technē*. The not very satisfactory defence to that is that those examples rejoin the main argument (from 639b21ff.) concerning types of necessity – which had been interrupted by the digression that introduces the contrast between the theoretical sciences and art. Option 2b avoids *that* difficulty and allows full weight to the two examples, health (the product of the art of medicine) and man (a case of something that comes to be *phusei*) but it in turn faces its embarrassment, which is that *phusikē* comes on one side of the contrast, *ta phusei* on the other, and yet this is supposed to be an account of how the former deals with the latter, indeed of the mode of necessity in question, indeed the mode of *demonstration* involved. However one might save the contrast where necessity is concerned, there seems no way of driving a wedge between the manner of the demonstration of *phusikē* and some manner of demonstration of *ta phusei*: yet manners of demonstration are what the passage is about, both in the ambiguous contrast and again where Aristotle speaks of the type of demonstration that cannot be joined to the eternal (640a6f.).

But I am not so much concerned here to engage in a comparative analysis of the merits and demerits of these rival lines of interpretation, as to see what follows for Aristotle's views on dem-

---

[17] See, for example, Düring 1961, and cf. Pellegrin 1982, pp. 164ff. (Pellegrin 1986, pp. 131ff.).
[18] See Kullmann 1974, pp. 13ff. There is a further brief discussion of some of the difficulties presented by this passage in Gotthelf 1987, pp. 197–8.

onstration *on any reading*. It is important to pay due attention to the fact that the passage is not *just* concerned with modes of necessity, the conditional and unconditional. This is what most commentators focus on (and most point out that there are further modalities of necessity not explicitly mentioned in this chapter). But Aristotle also wants to differentiate modes of *apodeixeis* and on *any* reading I believe we have to conclude that there is more to demonstration in the domain of physics than the model that is presented in *APo*.

Different readings, to be sure, give different ways of cashing out those differences. Take option 1, first, where the contrast is between physics on the one hand and the other theoretical sciences on the other. That has greater or less implications depending in turn on whether we take the next sentence, 640a3–4, which contrasts those inquiries where the starting-point is what *is* with those where it is what *will be*, as correlating exactly with the contrast in the disputed sentence. On that strong view, the contrast would be between the theoretical sciences that begin with what is, and physics here said to begin with what will be. On that reading physics would be understood as engaged in a manner of demonstration that has *nothing* in common with the theoretical sciences, and the prospect of even reconciling the physics with the *APo*. (let alone seeing the physics as trying to implement the *APo*. model) recede. There is, however, a weaker reading — still with option 1 — that at least preserves the prospect of compatibility. It is not necessary, and it may be quite undesirable, to take the inquiries that start from what is and those that begin with what will be as correlating exactly with the theoretical sciences and physics. Indeed if that had been Aristotle's meaning, he should have written *tais men, tēi de*, not *tois men, tois de* at 640a3. No doubt the theoretical sciences *are* limited to *to on* as their *archē*. But it is perfectly possible to see physics as concerned both with *on* and with *esomenon*. That still gives a contrast with the other sciences but a non-exclusive one. But so far as demonstration goes, it would still be the case that Aristotle is here insisting that there are *other* modes of *apodeixis* in physics besides those that it may share with the other theoretical sciences, those in which the *archē* is *to on*.

Something like that conclusion emerges also on the other two options I mentioned. Both 2a and 2b take it that the initial contrast is not *between* physics and the theoretical sciences, but between those two on the one hand and something else on the other (the products of *technē* in 2a: those and those of nature on 2b). But though that keeps physics and the theoretical sciences together at stage one of the argument, the later reference to a mode of demonstration that 'does not join the necessity to the eternal' – exemplified (as even 2a has to concede) by man as well as health – inevitably (re-)introduces *a* mode of demonstration that is quite different from the modes of demonstration discussed in *APo*.

We reach the crux of the matter. Even if an option 1 reading of the initial contrast is *not* adopted, the continuation of the passage commits Aristotle (on *any* view) to *some* new mode of demonstration dealing with man and exploring the conditional necessity of things *phusei*. This is a mode of *apodeixis* where you start from what will be, the *telos* – health or man – and work back to what had to be, or become, to produce that *telos*. The argument is deductive, but it does not deduce what follows from the starting-point in the sense of subsequent changes or processes, rather it deduces the antecedents of the *telos*. One might call it the deduction of the antecedents, though antecedents in a chronological or ontological not a logical sense.

PA 1 1 is no doubt not as transparent as one should like: the ideas are not even as worked out as the discussion of the practical syllogism in the ethics. But Aristotle's concern explicitly to point out that there is more than one mode of necessity, more than one mode of demonstration, cannot be dismissed as just some unfortunate anomaly. On the contrary, everyone must agree that the notion of conditional necessity plays a fundamental role throughout Aristotle's zoology. To those preoccupied with the notion of demonstration presented in *APo.*, at least in 1 1–4, it must come as a surprise to find Aristotle calling the deduction of the antecedents a mode of *apodeixis*: for there was no hint in those opening chapters of the possibility of a variety of modes of demonstration in one of which the condition of eternity is relaxed and the necessity in question is not unqualified but merely hypothetical. The theory of the *APo.* certainly does not prepare us for *that*. However if it is a surprise, we cannot just override his so doing. Not just because he says that there is this mode of *apodeixis* in a text devoted (as I said) to setting out the right method in zoology: but also because so far as his practice of argument goes, the mode of demonstration he here identifies tallies well enough with his own actual concern with exploring the relations between ends and their conditionally necessary antecedents in the body of the zoological treatises.

True those explorations do not usually come labelled explicitly as *apodeixeis*. But then very few arguments in the zoology are so labelled. As I remarked, the vocabulary of *apodeixis, apodeiknumi* is comparatively rare throughout the physical treatises, and in fact in *PA* itself the *only* two occasions when *apodeixis* is used are the two occurrences in *PA* I I. But the rarity of the terminology of *apodeixis* poses a *general* problem, not one that is specific to the question of the applicability of the idea of deduction of the antecedents. After all one (abstemious) reaction to that rarity would be to conclude that (whatever the expectations of a reader of *APo*. might be) in practice in the physical works Aristotle is very little concerned with *demonstration in any sense*. That may be *too* abstemious. But so far as the explicit usage of *apodeixis, apodeiknumi* goes, the lessons to be drawn from our analysis may be thought to be sobering enough.

To recapitulate and summarise my argument. We have, in the *Organon*, a powerful presentation of a strict notion of demonstration that has to meet very stringent conditions indeed. Moreover the demand, in the *Organon*, is that to be able to claim to *know* what is demonstrable you have to *have* the demonstration. Yet that is not the only theory of *apodeixis* we find in Aristotle. There is a second fully elaborated theory of rhetorical *apodeixis* in the *Rhetoric*. There are modes of *apodeixis* we encounter in the *Metaphysics* and elsewhere that do not meet all the conditions set out in the *APo*. We have not just straight, but also elenchtic demonstration, we have demonstrations that are more necessary, others that are laxer.

That already serves notice on us that we had better be careful when talking about demonstration in the physical treatises, to examine *what* mode or manner of demonstration is in question, and it is certainly not enough merely to gesture towards *some* interest in demonstration of *some* sort and use that as an indication that Aristotle is in business applying the model of *APo*. Of course our expectations are raised by the use of physical examples, as well as mathematical ones, in the *APo*. itself, even though many of the syllogisms, or demonstrations, in question use such unsatisfactory terms as 'near' or 'non-twinkling'. But whatever expectations these examples may generate, there is no substitute for the analysis of the physical treatises themselves, to determine Aristotle's actual practice.

But in the passages we have reviewed, the *nearest* Aristotle gets to an explicit recognition that *APo.* style demonstration *might be possible* in physics is the very text in *PA* I I which is principally concerned to show that there is *another* mode of demonstration, that of the deduction of the antecedents. On the weak reading of option I, where the contrast between physics and the other theoretical sciences is non-exclusive, the possibility remains open that in some cases, dealing with *to on*, physics too, like mathematics, has to do with what is eternal and with unconditional necessity. Yet elsewhere in all the other texts I mentioned reservations or limitations of one sort or another are stated or implied concerning the types of demonstrations that can be attempted in physics. We may have to settle for merely possible accounts: we may be engaged in dialectical arguments where the demonstrations in question are the sources of *aporiai* for opposing views: even the *archai* in play may be not indemonstrable primary premises, but principles of a quite different sort, namely efficient causes. And even in *PA* I I itself more than just demonstration of unconditional necessities is evidently in mind.

We cannot infer from these texts that Aristotle has abandoned the *APo.* model of demonstration entirely. That would be overinterpretation in one direction. But overinterpretation in the other direction is just as much to be avoided, I mean the presumption that because he refers from time to time to *apodeixis* in the physical treatises, the *apodeixis* in question has to be the notion we are familiar with from *APo.* It is clear from the evidence we have considered that when he speaks of *apodeixis* outside the *APo.* it is an open question how strict a notion is in play, and clear too that in the physical treatises he *often* has a looser sense in mind. Although he does not, in those treatises, set out a fully developed alternative to the *APo.* model, some relaxation of its conditions is undoubtedly often implied. Moreover it is not just that Aristotle's *practice* diverges from what the model of *APo.* lays down, but his explicit references to *method* indicate, I believe, his recognition that another *model* is needed.

That is only surprising if we assume Aristotle to have been quite inflexibly committed to a monolithic model of demonstration. But the theories of *APo.* are not just often *not* applied in practice in the physical treatises: in some cases they are not applic*able*. The main example of non-application is the lack of clearly set out syllogistic arguments. As for non-applicability, I have suggested elsewhere[19] that the univocity requirements of *APo*. go by the board in much of the *practice* of the physical investigations *and that had* Aristotle tried to alter the practice to meet the requirements that would have been a step in the wrong direction. But the idea that Aristotle may not just have needed to modify, but *saw* that he needed to modify, some aspects of the model to suit the actual problem-situations of physical inquiries, that idea is not just a conjecture. It gains positive support, I submit, from the analysis of his use of *apodeixis* that I have attempted.[20]

Postscript

Since this article was first written, the debate has continued (Lennox, Gotthelf, Kullmann, forthcoming) and to focus further discussion it may be useful to summarise where I see the main areas of agreement and disagreement. There are at least two main points where I believe it to be generally agreed that the practice of the physical treatises tallies well enough with the recommendations of *APo*. These relate first to the dependence of understanding on causal explanation (cf. above p. 8) and secondly to the asymmetry of causes and effects and to the corresponding linearity in the direction of causal reasoning. Valid deductive arguments are a necessary, but not a sufficient condition for correct explanations, which must proceed through the true explanantia (cf. p. 12).

The crux of ongoing disagreement relates first to the extent of the influence of other aspects of the ideal presented in *APo*. on

---

[19] Lloyd 1987, ch. 4.
[20] The text of this article is substantially that from which I delivered my lecture at Boston University in April 1990, and which had formed the basis of a seminar at Princeton the previous month. I received many constructive criticisms from those who participated in the discussions on those occasions or who have corresponded with me subsequently and I must express my warm thanks to Professors John Cleary, Myles Burnyeat, Michael Frede, David Furley, John Murdoch, Gisela Striker, and William Wians in particular. The revisions I have been able to make in the light of these comments and further reflection have been less substantial than I should have wished. This has been in part owing to the exigencies of the timetable of publication. Nor would it be appropriate for me within the conventions of this journal to modify my views at points taken up by Professor Wians in his reply. I hope, however, to be able to return to the major issues addressed on a subsequent occasion.

the physical treatises, in particular in the matter of the axiomatic-deductive style of demonstration set out in the opening chapters of *APo.* 1, and secondly to the interpretation of what is said about demonstration in such texts as those I have discussed in *Metaphysics* E 1 and *PA* 1 1.

So far as Aristotle's actual practice goes, the physical treatises do not set out their results in a fully articulated axiomatic-deductive system. But that by itself does not show that that was not the goal towards which Aristotle was working. Then so far as *Metaphysics* E 1 and *PA* 1 1 go, everyone must allow that he recognises some differences between different modes of demonstration in different contexts: the question then is, are these differences such as to call into question the applicability of the *APo.* ideal even as an ideal? Are the modes of reasoning that physics uses bound to fail by the strictest standards of axiomatic-deductive demonstration?

In my view, one of the great strengths of that ideal is indeed its uncompromising character. The indemonstrables must indeed be primary, self-evident, necessary truths. There can be no half measures about the axiomatic status of the axioms, including those specific to certain subject-areas. Equally the necessity required is unconditional.

Yet the very rigour of the criteria he sets up makes it all the harder to see how large areas of physics could ever – even in principle – fulfil them, to yield results that could eventually be presented in the form of a comprehensive axiomatic-deductive system. Moreover if we go by what we find in *Metaphysics* E 1 and *PA* 1 1, it seems that Aristotle there recognised as much himself, for he there allows laxer modes of demonstration, more or less necessary ones, or ones based on a necessity that is conditional, not the absolute necessity that belongs to what is eternal.

But once the pluralism of Aristotle's views on demonstration is given full weight – a pluralism that stretches after all as far as the *Rhetoric* – that removes one motive for interpreting the *PA in terms of APo.* It will not do merely to observe that Aristotle does not deny demonstration in physics: we have to specify *what kind* of demonstration he has in mind. *PA* 1 1 has, rather, to be taken on its *own* terms, as allowing a mode of demonstration that does not conform to the criteria of *APo.*, most notably in that it does not take as its starting-points the primary indemonstrables identified in *APo.* 1 2 (above p. 12), but uses end-products as its starting-points and works back to their conditionally necessary antecedents. While the reasoning is deductive, the nature of the premises, the mode of necessity and the goal of the inquiry all differ from those in view in the opening chapters of *APo.* 1.

The analyses we shall offer of other aspects of Aristotle's work in the studies that follow will provide further examples that suggest a similar moral, underlining the variety of theories and practices in his scientific methodology. In particular we shall examine cases of his tolerance of polyvalence (chapter 4) and his explicit recognition that the *per genus et differentiam* model of definition is not applicable to some of his key metaphysical and physical terms, which have to be analysed, rather, in terms of his notions of analogical predication and focal meaning (chapter 7). We shall have further occasion, too, to consider the problems posed by any departure from strict univocity (chapter 7), including metaphor (chapter 10), for deductive validity itself.

CHAPTER 2

The relationship of psychology to zoology*

In prominent texts in the *de Anima* and his zoological treatises Aristotle stresses the importance of the study of soul for the student of nature (*phusikos*). The aim of the present chapter[1] is twofold: first, to investigate the extent to which his zoological researches were in fact guided or influenced by his general psychological[2] theory and by his specific psychological doctrines; and secondly, to explore the match or mismatch between the results of his zoological investigations and his general position on such questions as definition, essence, form, and matter. My thesis will be, on the first count, that the psychology does indeed provide what is, from some points of view, the major articulating framework for his zoology, though it is a framework that leaves plenty of scope for the introduction of supplementary material not immediately and maybe not even ultimately geared to resolving questions connected with his main psychological interests. Moreover, I shall argue that at certain key points in his zoology his specific psychological doctrines strongly influence his interpretation of the biological phenomena. Then on the second count I shall try to clarify what I take to be deep-seated tensions between his account of living creatures and his doctrines of definition and of form. For the issue of the credibility – for us – of his doctrine of *psuchē* it may be important to pose the prior

---

* This article is reprinted in its original form from *Essays on Aristotle's De Anima*, edd. M. C. Nussbaum and A. O. Rorty (Oxford, 1992), pp. 147–67, with minor bibliographical corrections and the addition of some supplementary notes, indicated with an asterisk, and a brief postscript.
[1] Among important recent discussions of topics broached in this chapter I may mention especially Balme 1961/1975, 1962a, 1987a, 1987b, Sorabji 1974/1979, Barnes 1971–2/1979, Ackrill 1972–3/1979, Bolton 1978, Lennox 1980, 1987a, 1987b, Gotthelf 1985, 1987, Frede 1985, Kosman 1987, Burnyeat 1992.
[2] Throughout this chapter I use the term 'psychological' not in our usual English sense, but in that of Aristotle's doctrine of *psuchē*.

question: was it credible even for Aristotle? The anti-functionalists are undeniably right to insist that, in the case of living creatures, there is no alternative to the matter they happen to possess. But the functionalists have a point, that the expectation generated by certain of his explicit statements on form and definition is that there should have been. There are problems here not just for our latter-day evaluation of Aristotelian philosophy of mind, but also internal to Aristotelianism itself. The hesitations I think we can detect on Aristotle's part can be seen as evidence of *his* realisation of some of the problems. This is a case where we should resist trying to *unify* his doctrines beyond a certain point, though I am aware that that is always a difficult line of interpretation for which to argue.

First, however, some of the programmatic statements should be set out. In the opening chapter of *de Anima*, as he embarks on the inquiry concerning soul, he makes two important remarks. The first, 402a4ff., reveals the strategic importance of psychology. It is agreed, he says, that knowledge of the soul contributes much to 'the truth as a whole', and especially to the understanding of nature, and this statement is immediately explained with the remark that soul is, as it were, a principle for animals. Secondly, he stresses that the inquiry should not be limited to *human* soul, though that is the way his predecessors had treated the problem (402b3ff.). As is well known, that principle is repeatedly put into practice in the *de Anima* and the *Parva Naturalia*, not to speak of what we think of as the more purely zoological treatises themselves. Regular consideration is given to such questions as the different sense-faculties that different species of animals possess, whether they sleep, have imaginations, memory, and so on. This already gives a biological or zoological orientation to his psychological discussions. Conversely, when he considers zoological methods in the opening chapter of *de Partibus Animalium*, he notes at 641a17ff. that the student of nature must treat of the soul as being the form and the substance (*ousia*) of the animal. However, *phusikē* is not concerned with the whole soul (including reason, *nous*), for if that had been the case – he adds in a further revealing remark, *PA* 641a34ff. – then there would have been nothing to philosophy over and above *phusikē*, the study of nature.

One of the major themes of Aristotle's psychology, indeed its chief foundation, is, of course, that soul is to body as form is to matter, or more strictly as first actuality is to matter. This provides him with his answer to the question of how soul and body are related. Indeed he feels able to claim, notoriously, at 412b6ff., that there is no need to inquire whether the soul and the body are one, just as there is no such need in the case of the wax and the shape made in it (e.g., by a seal), nor in general in relation to the matter of each thing and that of which it is the matter. But while this shows great apparent confidence in the unity of the living creature, aspects of which can be considered under the separate rubrics of 'soul' and 'body', it is nevertheless the case that Aristotle still insists rather sharply on the differences between these two, not least because the soul is, in a sense, that for which the body exists (e.g. *PA* 645b19). Repeatedly in *de Partibus Animalium* I especially Aristotle puts it that the chief topic of investigation for the student of nature is soul and the composite whole, rather than the body or the matter (e.g. *PA* 641a29ff., 645a30ff.). Yet while in the single case of the activity of reason we have a clear instance of an activity of the soul that does not involve the body (for reason, as is well known, is not the activity of any bodily organ in Aristotle's view), elsewhere the *distinction* between soul and body becomes more problematic the more Aristotle stresses their *interdependence*. Thus he recognises that perception, for instance, is not an *idion* (property) of the soul, no more is it one of the body (*Somn.* 454a7ff.). But just as, in perception, the faculty of the soul involved does not operate on its own (as happens in the case of reasoning), so conversely on the body side of the equation it is not just any body, any inert physical stuff, that can see or hear: in particular, as he never tires of saying, a dead eye is an eye only in name.[3]

Aristotle's provisional resolution of those difficulties proceeds partly via his distinction between potentiality and actuality (the soul being the first actuality of a natural body that potentially has life, *de An.* 412a27f., or as he puts it at 412b5f., of a natural instrumental body) and partly via his recognition that some faculties are 'common to soul and body', as at *Sens.* 436a6ff. for instance, where he specifies quite a range of such items including perception, memory, anger, desire, pleasure, and pain. But while in principle a recommendation to study on the one hand the material or physical aspects of vital activities, and on the other what makes them the vital activities they are, may look unprob-

[3] See, e.g. *PA* 643a24ff., *GA* 726b22ff., 734b24ff., 735a6ff., 741a9ff.

lematic enough, in practice the correlativity of form and matter, actuality and potentiality, ensures that they cannot be treated independently of one another. How far that in turn threatens the crisp application, to zoological phenomena, of some of Aristotle's recommendations concerning definition (particularly when that is said to be of the form) is a question that will occupy us in Section II of this chapter.

I

The expectation generated by the programmatic statements that Aristotle makes in the *de Anima* and the *de Partibus Animalium* is that zoology will be largely, though to be sure not exclusively, devoted to a consideration of *psuchē*. Given that zoology is a study of living creatures and that what makes them the living creatures they are falls under the rubric of *psuchē*, that should occasion no surprise (however much later uses of the terms 'psychology' and 'biology' make it anything but obvious why the first should underpin the second). But if it is one thing for Aristotle to set out his programme for the correct method for studying animals, it is another for him to implement it. How far do the enormously rich and wide-ranging zoological treatises conform to a plan that tallies with his expression of a primary concern with *psuchē*? Do his particular psychological theories influence his zoological explanations, and if so how?

Part of the answer to the first question is straightforward enough. Obviously, whenever he is dealing with an instrumental part that is directly concerned with one of the major faculties of the soul identified in the *de Anima*, Aristotle cannot fail to bear in mind precisely that *that* is the function that the part serves, and he will indeed see the activities in question as the final causes of the parts. This is the case, for instance, of his various discussions of the organs of locomotion, especially of the detailed account, in the *de Incessu Animalium*, of the different modalities of locomotion, for example of the differences between birds' wings and insects' wings and of the different ways in which the front and back legs of quadrupeds bend. Again perception has a fundamental role in his general definition of animal, being, of course, the faculty that, in Aristotle's view,* distinguishes an animal from a plant; and, as

---

* That is, his usual view: whether he allows exceptions will be discussed below in ch. 3.

already remarked, the questions of which animals have which senses, and the varieties in the sense-organs, naturally receive very careful discussion.

The various organs that serve the complex functions of the first faculty of soul, *threptikē*, covering both nutrition and reproduction, are a third clear example with interesting and complex ramifications. Among the instrumental parts that serve the function of nutrition are mouth, teeth, lips, tongue, stomach, liver, omentum, mesentery, and the whole digestive tract – though several of these parts have other functions as well, the tongue for taste, the teeth for defence, and tongue, teeth, and lips for voice and speech, for instance.[4] Moreover, since nutrition necessarily involves not just the intake of food but also the excretion of residues, the parts that serve the latter function too must be included, the kidneys and bladder for example. In general, as he often states in the *de Generatione Animalium*, there is no part of the living body that does not have soul.[5] But Aristotle is particularly exercised, in his account of reproduction, to specify *how* the soul is present in the *seed*, distinguishing the nutritive soul, which is present already potentially in the seed (and comes into operation as soon as the seed draws nourishment to itself, *GA* 736b8ff.), from the perceptive soul, which is supplied by the male parent and is present, again potentially, only at the point when a new animal is recognisable as such.[6] Again, his whole theory of the roles of male and female in generation revolves round his idea that the former provides the form and the moving cause, the latter the matter, a doctrine expressed at *GA* 738b25ff. in terms of the male supplying the *soul*, the female the body.[7]

---

[4] See, e.g. *PA* 660a19ff., 22ff., 35ff. on the tongue, *PA* 659b30ff. on the lips, and *PA* 661b1ff., 13ff. and *GA* 788b3ff. on the teeth.

[5] See, e.g. *GA* 726b22ff., 734b24ff., 735a6ff., 741a23ff. However the controlling principle is located in the heart or analogous part, e.g. *Juv.* 467b14ff., *MA* 703a34ff.

[6] See, e.g. *GA* 735a4ff., 16ff., 736a35ff., b1ff., 738b25ff., 757b15ff.: on how reason is acquired or transmitted, see *GA* 736b5ff., 737a7ff. See Code 1987.

[7] Cf. *GA* 737a27ff., where the female *katamēnia* are said to lack the *archē* of the soul. In *GA* IV 3, however, when dealing with the likenesses of offspring to parents (to the mother as well as the father, and to grandparents on both sides) Aristotle talks of the movements in the seed that are derived from both parents (e.g. 768a10ff.: contrast, now, the interpretation of *GA* IV 3 offered by Cooper 1988), and in general, while in some texts the differences between male and female contributions to reproduction are stressed, in others the emphasis is on the point that the female *katamēnia* are *analogous* to the semen in males (e.g. *GA* 727a2ff.) and are indeed seed, even if not pure, not fully concocted, and 'in need of elaboration', e.g. *GA* 728a26ff.

Clearly, when dealing with the instrumental parts or the sense-organs Aristotle *must* pay due attention to the faculties of the soul that they serve. But one might suppose that the relevance of his general and particular psychological theories to his account of the simpler, uniform parts of living creatures would be only a very limited one. However, in several instances that is certainly very far from being the case. Two examples will serve to make the point, his accounts of flesh and of blood.

As already remarked, the faculty that distinguishes animals from plants, in Aristotle's view, is perception.* To be more precise, what all animals possess is the primary mode of perception, namely touch.[8] However, what serves as the organ or more strictly the medium of touch is flesh, and in his account of flesh in *PA* 653b19ff., it is this that provides his initial and primary focus of interest. He is absolutely clear about the fundamental importance of flesh for every kind of living creature. Indeed, one of his ways of distinguishing between two of the main families of bloodless animals, the crustacea and the cephalopods, is on the question of whether their fleshy parts are on the inside or the outside of their bodies.[9] But instead of considering flesh from the point of view of the musculature, for instance, Aristotle deems its essential role to be as the medium of touch.

Moreover, this distinctive *cognitive* role of flesh has far-reaching repercussions on the account Aristotle offers of a wide variety of other uniform parts as well. At *PA* 653b30ff. he puts it that 'everything else' is for the sake of the flesh (or more strictly for the perceptive faculty of which it is the medium) and he specifies 'bones and skin and tendons and blood-vessels, and again hair and all kinds of nails and so on', explaining that, for example, the bones are devised for the preservation or protection of the soft, fleshy parts, and that in animals that have no bones some analogous part, such as spine or cartilage, plays a similar function.[10] Of course, Aristotle does not mean to suggest that there is no more to the function of the bones and the other parts he mentions than their being, in some sense, 'for the sake of' flesh. In particular he does not believe that mentioning that function absolves him from

---

[8] See, e.g. *Sens.* 436b10ff., *Juv.* 467b23ff., and the other texts cited below at n. 69.
[9] See, e.g. *HA* 523b2ff., 5ff., cf. *PA* 653b36ff.
[10] See, e.g. *PA* 653b35ff., and see also *PA* 654b27ff.
* Cf. however below, ch. 3.

the obligation of giving detailed descriptions and explanations of those other uniform parts, the courses of the blood-vessels, the specific function of hair on different parts of the body, the different arrangements of bones in different species of animals, and much else beside. Yet evidently flesh, in his view, has a primacy, among the uniform parts, and the reason for that is clear, namely the role it plays in the primary mode of perception.[11]

While flesh is said to be an *archē* or principle for the body as a whole in virtue of its role in the faculty of touch, blood is described as in a sense the matter for the whole of the body.[12] At first sight that might make it appear that the role of blood would be limited to accounts where Aristotle has the material cause in view and so be of less importance where formal, final, and efficient causes are concerned, that is to say where the soul, acting in those capacities, is. Yet that is far from being the case. First the blood itself certainly serves a psychic end and a fundamental one: its final cause is nutrition, *PA* 650b2ff., 12f.,[13] and it is, potentially, the body or flesh, *PA* 668a25ff. The network of blood-vessels serves other purposes as well, being compared to a framework for the rest of the body, *PA* 668b24ff., binding together the front and the back;[14] but their fundamental function is to nourish the body, and it is for that reason that they permeate every part of the body.[15]

But there is far more to the role of blood than just nutrition, since in Aristotle's view it contributes directly to a whole range of other functions of the soul. The nature of the blood, he says at *PA* 651a12ff., is responsible for many things both in respect of the *character* of animals and in respect of *perception*, in a chapter which has offered a variety of suggestions about how the intelligence, acuteness of perception, courage, and timidity of different species of animals are to be correlated with the quality of the blood they

---

[11] See further *de An.* 423a13ff., b26, 426b15, *PA* 647a19ff., 656b35f.
[12] See e.g. *PA* 668a5f., 25ff., *GA* 751a34ff.
[13] See further, e.g. *GA* 726b1f., 740a21f., b3, *Somn.* 456a34ff., *Resp.* 474b3ff., *PA* 652a6f., 678a6ff.
[14] Cf. *GA* 743a1ff.: yet at *PA* 670a8ff. he suggests that the lower viscera serve to anchor the blood-vessels.
[15] See, e.g. *PA* 668a4ff., 11f. At *PA* 668a13ff. the blood-vessels are compared with a network of irrigation channels, and at 668a16ff. with material laid out along the foundations of a house.

possess, its heat, purity, thinness, whether it contains fibres, and so on.[16] Again, as is the case with his theory of flesh, the account given of blood has far-reaching repercussions, since he treats a number of other uniform parts, lard, suet, marrow, and so on as kinds of blood or derivatives from it.[17]

The two cases we have considered, flesh and blood, illustrate how, even at the level of the fundamental uniform parts that form the material of the living body, Aristotle's account is concerned with far more than just their role as stuff and in each case he develops distinctive theories of the vital functions that these parts serve. But we can go further still. The primary simple bodies in Aristotle's element theory are, of course, earth, water, air, and fire, considered as combinations of the primary pairs of opposites, hot and cold, wet and dry. But when he comes to consider those opposites in *PA* II he leaves us in no doubt that it is not just as inert material that they are important. The very reason why it is essential to get clear about hot and cold, wet and dry – topics on which there had been so much dispute and confusion in the past – is that 'it seems clear that these are responsible practically for death and life, and again for sleep and waking, for maturity and old age, and disease and health'.[18] They are responsible, of course, in the first instance as material causes.[19] However, we have to be clear that these primary qualities are not treated in the zoology as merely physical in our sense, where that is contrasted with biological.

Thus of the pair wet and dry he tells us at *GA* 733a11f. that wet is particularly associated with, or productive of, life (*zōtikon*), while the dry is 'furthest from what has soul' (*to empsuchon*).[20] More strikingly still, the kind of heat that he refers to repeatedly in his explanation of generation in *GA* is explicitly described not just as what is appropriate to living creatures, their proper or own (*oikeion*),

---

[16] See esp. *PA* 647b31ff., 648a2ff., 650b18ff., *Somn.* 458a13ff., *Resp.* 477a18ff. At *PA* 667a11ff. Aristotle further suggests that differences in the size and texture of the heart contribute also to character, *ēthē*.
[17] See, e.g. *PA* 651a20ff., b20ff., and cf. *GA* 726b9ff. on seed. Furthermore, at *PA* 673b26ff. he suggests that the liver's *telos* resides chiefly in the blood, and cf. *PA* 674a6ff., on viscera that are for the sake of the blood-vessels.
[18] *PA* 648b4ff., cf. *GA* 784a32ff. on ageing.
[19] See e.g. *Long.* 466a18ff.
[20] Cf. also *Long.* 466b21f., *HA* 489a20f., *PA* 647b26f.

heat,[21] but as the vital (*psuchikon*), heat.[22] Of course the relation between vital heat and just ordinary heat raises plenty of difficult questions, some of which the Aristotelian discussions leave rather unresolved, as for example whether we are to think of vital heat as indeed a separate kind of hot or as the hot working in a particular fashion.[23] But whatever answer we offer to that question, it is abundantly clear from the very characterisation of it as *vital* that in this context Aristotle has in mind far more than a merely physical quality (again, in *our* sense).* What little Aristotle has to say on the subject of *pneuma* is notoriously obscure and has occasioned protracted scholarly debate:[24] yet some of the perplexing features of that doctrine – the role of *pneuma* as the material instrument of psychic activities – are shared, to a greater or less degree, by the primary qualities. On the one hand Aristotle is aware of the dangers of hylozoism and explicitly criticises some of his predecessors for their failure to pay due attention to the differences between the animate and the inanimate.[25] On the other, his own view of the material substrate of animate beings incorporates the crucial assumption that the matter in question is not just inert stuff: as the definition of soul at *de An.* 412a20f. puts it, what the soul is a first actuality *of* must itself possess the potentiality for life – a potentiality exemplified not just by the instrumental parts, but to a lesser degree by the uniform parts of which they consist, and indeed by *their* material components, even when these are analysed in such apparently purely physicalist terms as the hot, the cold, the wet, and the dry:[26] that appearance, we may say, is rather deceptive.

---

[21] See, for example, *GA* 784a35f., b5, 786a20f.
[22] See, for example, *GA* 732a18f., 739a11, 752a2f., 755a20, 762a20, cf *zōtikē thermotēs* at *Resp.* 473a9f. and the frequent references to natural (*phusikē*) heat, e.g. *GA* 732b32, 766a35, b34, 783b30, 784b26, 786a11, *PA* 650a14, *Juv.* 469b8ff., 470a19ff.
[23] At *GA* 736b34ff. the 'so-called hot' is contrasted with fire and said to have the nature of *pneuma* in it, which in turn is said to be 'analogous to the element of the stars'. Again at *PA* 652b7ff. Aristotle criticises as crude the notion that the soul of an animal is fire or some such *dunamis*, though he allows that it might be said to subsist in some such body, and at *GA* 762a18ff., discussing how animals and plants are formed in earth and the moist, he says that there is water in earth, and *pneuma* in water, and vital heat in all *pneuma*, concluding that 'in a way' (*tropon tina*) everything is full of soul.
[24] See, e.g. Rüsche 1930, Peck 1943, 576ff., Verbeke 1945, 1978, Nussbaum 1978.
* Leaving the food in the sun to warm it does not digest it. No more does any or every application of heat concoct (that is fertilise) the menses. Heat is, to be sure, a necessary condition of both digestion and concoction, as also of complex processes, such as cooking, that take place outside the living body. But Aristotle recognises that in all such cases the heat has to be appropriate to the end in view and have the right capacity to bring

Of course the extent to which Aristotle's account of parts of the body proceeds by reference to the faculties of the soul varies a great deal. That feature of his approach is at its most prominent in such a case as the heart, seen by Aristotle as the controlling principle of the living creature,[27] and in particular as the centre of the faculties of locomotion, of perception, and of nutrition/ generation – a set of theses for which he argues in some detail, and in the case of perception, at least, in the face of some difficult anatomical facts.[28] But even when a part is not directly linked to a specific faculty of the soul it may, and generally does, serve such faculties indirectly or the general well-being of the living creature as a whole. Thus his view of the brain and the lungs is that in different ways they both act primarily to balance the heat in the region round the heart, and so indirectly help to ensure its activities and the life of the animal as a whole.[29] To be sure, some parts, he explicitly insists, have no final causes. The useless residues (unlike the useful ones, such as semen) are just that, the end-products of such processes as digestion. One example is bile, where interestingly enough he expressly contradicts the view of those who had ascribed a role to it in perception,[30] and where he states the general doctrine that while some parts have a purpose, others arise of necessity as a consequence of these, *PA* 677a15ff. Yet even here though the useless residues serve no good, the account he offers of how they arise, of necessity, in the body, refers to those processes of which they are the by-product; and those processes themselves,

---

it about (e.g. *GA* 743a26–34). We shall be reverting to the problems of concoction and generation in chh. 4 and 5.

[25] See, esp. *de An.* 411a7ff., 14ff. As is well known, however, Aristotle himself admits that the dividing line between what does, and what does not, have *psuchē* is hard to determine and that nature passes in continuous sequence from the inanimate to the animate, *HA* 588b4ff., *PA* 681a12ff.*

[26] Value-judgements are also in play in some of Aristotle's uses of these four primary opposites, most notably in the frequent association of what is hotter with what is nobler; see e.g. *Resp.* 477a16ff., *GA* 732b31ff., 733a33ff.

[27] See e.g. *Juv.* 469a10ff., *PA* 647a24ff.

[28] I detailed these in my (1978) 1991, 238ff. especially those connected with Aristotle's claims that the senses of taste and touch evidently extend to the heart and so the other senses necessarily do so too, *Juv.* 469a12ff.

[29] See esp. *PA* 652a24ff., b6ff., 656a19ff., *GA* 743b29ff. on the brain and *PA* 668b33ff. on the lungs.

[30] See *PA* 676b22ff., where Aristotle appears to have Plato, *Timaeus* 71aff. in mind. Cf. Aristotle's criticisms of the view that the brain is the seat of perception at *PA* 656a15ff., though certain senses are located in the head, 686a8ff.

* Cf. below, ch. 3.

digestion for example, take one inevitably back to the functions of the soul.

Two final examples may be given to illustrate the pervasiveness of psychological considerations in Aristotle's zoological theories. In the seemingly unpromising case of the diaphragm, which acts as a membrane separating the upper from the lower viscera, the particular function that Aristotle sees it as serving is to ensure that the perceptive soul (in the heart) is not affected too rapidly by the exhalations that arise from the processes of digestion.[31] Nature likes to keep the nobler parts apart from the less noble, when she can, but 'nobler' in this case is the perceptive faculty of soul and the 'less noble' (by implication) the nutritive function.

Then it is particularly remarkable that even in the case of the three pairs of dimensions, up–down, right–left, and front–back, in living creatures, Aristotle's account is essentially psychological in orientation. These three pairs are not just spatial differentiations, nor are they value-neutral. They are each defined, so far as the animal body goes, in terms of a particular faculty of the soul: thus up is the direction of growth and that from which nourishment is taken in,[32] right is the principle of beginning of movement,[33] and front, the principle of perception.[34] These theories incorporate value-judgements – for the three terms that are principles are superior to their contraries[35] – and they are invoked in a whole series of detailed explanations of anatomical and zoological facts, such as the relative positions of the windpipe and oesophagus, those of the two kidneys, and the position of the heart, down to such questions as why in general the right claw of crabs and crayfish is bigger than the left.[36]

Thus far we have been concerned mainly with the theories and explanations advanced in *PA*, *IA*, and *GA*, and the question that

---

[31] See *PA* 672b14ff.: but he denies the view that the *phrenes* participate in thinking, *PA* 672b31ff.

[32] See e.g. *IA* 705a32ff. This doctrine leads Aristotle to the conclusion that plants are – functionally – 'upside-down' since they take in their nourishment through their roots, *PA* 686b31ff., *IA* 705b6, cf. *PA* 683b18ff. on the Testacea.

[33] See e.g. *IA* 705b29ff.

[34] See e.g. *IA* 705b8ff.

[35] See esp. *IA* 706b12f., *PA* 665b22ff., and cf. Lloyd 1966, 52ff.

[36] See Lloyd 1966, 52ff. which discusses also the particular difficulty (recognised by Aristotle) presented by the fact that the human heart is on the left, *PA* 665b18ff., 666b6ff.: cf. Byl 1968, 1980, who stresses rather the symbolic importance of the heart's being in the centre.

Relationship of psychology to zoology 49

now arises is how far similar considerations are at work also in the more purely descriptive inquiry in *HA*.[37] *HA* has often been thought of as a fairly unsystematic, diffuse, and at points repetitive presentation of primary data. Yet, as has been argued most recently by Lennox and was indeed already clearly shown in Peck's analytic table of contents,[38] the overall plan for the study of the differentiae of animals that is set out in *HA* I is adhered to – on the whole – quite closely, at least so far as *HA* I–IX go.[39]

At *HA* 487a11ff. Aristotle identifies four modes of differentiation – in respect, namely, of lives (*bioi*), activities (*praxeis*), characters (*ēthē*), and parts (*moria*) – and after the preliminary sketch (*tupōi*) of these in *HA* I 1–6, the whole of *HA* I 7–IV. 7 deals with parts, books V–VIII discuss activities and lives, and book IX considers characters, though of course there are digressions, within each of these main sections, that do not conform to this simplified schema. But the first point that may strike one concerning the particular differentiae that Aristotle uses to organise his material is that three of the four relate directly and primarily to the soul. It may appear to some modern commentators to be rather strange that his zoology should extend to the study of animal *characters*, though I have already remarked on the references to the intelligence and courage of different species in Aristotle's theory correlating these with the qualities of their blood.[40] Moreover, the nutritive and locomotive faculties of the soul bulk large in his account of *activities* and *lives*, when he discusses, for example, where different species of animals feed, whether they are swimmers or fliers or go on land, though there is, of course, much else to his discussion of those differentiae, including other primarily psychological factors such as whether a species is social (*politikon*) and whether it lives in groups or is solitary.[41]

Moreover, in his preliminary account of the fourth main type of differentia, the parts, at *HA* 488b29ff., he focuses first on the 'most

---

[37] This is not to deny that *HA* has its theoretical concerns and assumptions: cf. Balme 1987*a*, 88f., Lennox 1991.
[38] See Lennox 1991, and cf. Peck 1965, xciff., xcivff.
[39] *HA* X is anomalous and its authenticity (as also that of the other later books, 1–9) has been doubted: see, however, Balme 1985.
[40] See above on *PA* 651a12ff., the conclusion of the discussion in *PA* II 2ff. I discussed aspects of Aristotle's views on the characters of animals (e.g. *HA* 488b13ff.) in my (1983), 18ff.
[41] See especially *HA* 487b34ff.

necessary' parts of an animal – namely, first those to do with nutrition, the intake and storing of food, and the excretion of residues, and then again those to do with reproduction – whereas perception, especially touch, and differences in the organs of locomotion figure prominently (though not exclusively) in the subsequent preliminary list. This doctrine of the necessary parts of an animal (that is, those that correspond to various essential vital functions), plays, as I tried to show some years ago, an important heuristic role.[42] At least in the detailed analysis of the internal parts of the bloodless animals, in *HA* IV 1–7, Aristotle is particularly concerned to identify what serves the three basic functions, (i) the intake of food, (ii) the excretion of residue, and (iii) the control of the vital functions as a whole – which he expects to find and usually does find in the centre of the animal.[43]

Thus he regularly considers such questions as the position of the mouth, the presence or absence of teeth and tongue or analogous organs, the position and nature of the stomach and gut, as also the reproductive organs and the differences between males and females.[44] A series of passages shows that he actively considered whether or not certain lower groups of animals produce residue and attempted to identify and trace the excretory vent. But while the whole course of the alimentary canal is thoroughly discussed in connection with each of the bloodless groups, he has little or nothing to say about the brain or about the respiratory (he would say refrigeratory) system. He has of course no idea of the nervous system, though his conception of the role of the heart as the controlling principle of the major vital functions answers the question of where the ruling principle of the animal is located. But although he recognises that some creatures continue to live, even when divided – and they have *several* principles of life[45] – his expectation is that normally there will be just one such centre and indeed that it will be (as the heart is) in the middle of the body. Naturally, then, interpreting the function of the brain in the larger, blooded animals as one of refrigeration, he also misses the true role of the analogous parts in such creatures as the cephalopods, where the influence of his expectations concerning the func-

[42] See Lloyd 1979, pp. 213ff., on *PA* 655b29ff., *Juv.* 468a13ff. especially.
[43] See esp. *PA* 681b33ff., *Juv.* 467b28ff.
[44] The evidence is set out in detail in Lloyd 1979, 213f., nn. 438–41.
[45] See e.g. *PA* 682a1ff.

tion of the heart dominates his account of the central internal parts.

We may now attempt to take stock of this first section of our analysis. Given that Aristotle's doctrine of *psuchē* is a doctrine of the vital faculties of living creatures, and given further his explicit insistence on the importance of *psuchē* for the student of nature as a whole, we should expect – and we actually find – that both his general doctrine of soul and his specific theories distinguishing its various faculties exercise a profound influence throughout the zoological treatises. Of course those psychological theories are not the sole preoccupation of his work in zoology: other concerns, some of a strategic nature, can be identified. One that I have discussed elsewhere is his use of human beings as a model for other animals, a model to which they aspire and from which they deviate to a greater or lesser degree.[46] The framework provided by his schema of causation, formal, final, efficient, and material causes, or more simply the contrast between what is for the sake of something and the necessary, broadly coincides with his distinction between soul and body, but enables questions to be raised that a mere dichotomy between soul and body might obscure. The notion that Aristotle is further particularly concerned to establish correlations in order eventually to provide demonstrations of zoological propositions, has been argued with some force in recent years,[47] and other strategic and tactical aims for stretches of his zoology could be suggested. Moreover, many detailed descriptions of the parts of animals, and of animal behaviour, exhibit no specific direct influence from his particular psychological theories, whatever may be the indirect effects of his overall concern with soul.

On the other hand those theories are clearly invoked in contexts that extend far beyond his discussion of the instrumental parts or sense-organs that directly serve the main faculties of the soul – nutrition/generation, perception, locomotion. His account of two particularly important uniform parts, flesh and blood, ascribes to them a role in more than just the processes of nutrition, in cognition in the one case, and in character and intelligence in the other. The doctrine that the matter of the living creature has a poten-

---

[46] See Lloyd 1983, pt 1, §3, pp. 26ff. with references to previous discussions on p. 26 n. 56.
[47] See esp. Lennox 1987b.*
* But cf. above ch. 1.

tiality for life applies not just to instrumental, not just to uniform, parts, for it also extends as far as the primary qualities, at least when he invokes vital heat and the connection between wet and life. Even spatial dimensions, in the animal kingdom, are defined in terms of faculties of soul. The theory of the essential parts of the animal, corresponding to vital faculties, provides an important heuristic tool and an articulating schema for much of his zoological work. In such contexts we may talk not just of a general background of psychological interests – provided by the doctrine that soul is the form and final cause of the living creature – but also of specific influences from the detailed application to zoology of his particular psychological theories.

II

Having considered aspects of the relevance of the psychology to the zoology, we may now turn to discuss how the points we have made may relate to the interpretation of Aristotle's general theories on such topics as definition, essence, form, and matter. Of course what is said on those topics outside the zoology is highly complex and disputed: my aim here is to tackle some applications of those notions in the zoology, which provides, after all, our most extensive evidence of Aristotle at work in natural philosophy.

On the question of definition, first, some further preliminaries are necessary. Outside the zoology, as is well known, we may distinguish between what we may call a narrower view of definition and a broader one. According to the broader one, to give an account of a composite *sunolon*, both matter and form must be included. When something is a 'this in a this' (*tode en tōide*), then just to give an account of the form or essence is inadequate, for the account[48] of such an item should include matter as well. This is, moreover, regularly illustrated with reference to natural objects, *ta phusika*, including the parts of living creatures.[49] Indeed, since the *dia ti* question can be asked in respect to all four causes, the account of what a thing is may include all four.[50]

On the other hand there is also a narrower view of definition

---

[48] Aristotle often uses the term *logos*, but sometimes *horismos/horizesthai*, when such an account is in question: see e.g. *Metaph.* 1025b30ff., cf. 1064a21ff.
[49] See esp. *Metaph.* 1025b34ff., *de An.* 429b13f., see Balme 1987b, 306ff.
[50] See esp. *Ph.* 194b16ff., where the four causes correspond to four types of *dia ti* question.

according to which it relates to form or essence and matter is excluded. We should be careful to distinguish, as Michael Frede has recently done in a useful discussion,[51] between two possible ways of taking the general dictum that definition is of the form (or essence). This might correspond either to the (strong) view that no definition of a *sunolon* is possible: the only legitimate objects of definition are forms or essences. Or it could be the weaker view that the only way in which a definition of a *sunolon* can be given is *by specifying* its form/essence: but it is perfectly possible to do that and so in that sense a definition of a *sunolon* is not ruled out. But on both the weak and the strong view the definitions that are given do not include any reference to matter.

What the *Metaphysics* has to say about the definitions of animal and man is controversial, and I shall have to be brief on the undeniably complex questions of the interpretation of Z10–11 in particular, for the sake of getting to evaluate the actual practice in the zoology. Those two chapters introduce, famously, remarks to the effect that while the soul of animals is the essence, 'if one is to define each part well, one will not define it without its function (*ergon*), which does not belong without perception' (1035b16ff.); and again that one should not 'do away with matter' (1036b22ff.); and 'for an animal is something capable of perceiving[52] and it is not possible to define it without movement[53] and so not without the parts' being disposed in a certain way' (1036b28ff.). Quite *what* concessions are being made in those passages is untransparent, but it is essential to notice that when Aristotle comes to sum up the

---

[51] Frede (1990).
[52] At *Metaph*. 1036b28 the MSS and editions have *aisthēton*, perceptible, but Frede–Patzig 1988, ii, 210f., plausibly conjecture that we should read *aisthētikon*, capable of perceiving, which certainly gives a better sense.
[53] As Frede, 1990, points out (cf. Frede–Patzig 1988, ii. pp. 211ff.), this could be taken in two ways, either that it is not possible to define animal without reference to motion, or that it is not possible to do so without making it clear that the animal is in motion. Cf. Balme 1987b, 302ff., who has recently argued that the earlier chapters of *Metaph*. Z (esp. 5–6, 10–11) state an *aporia* that is only resolved in *Metaph*. Z17 and H6. However those two chapters address the problem of the unity of the genus and differentiae. Z10–11 mainly address a rather different issue, namely the unity of the constituents of the composite whole, where the parts in question are form or essence, and matter. Moreover, *after* a brief allusion to the problem of the unity of the parts of the definition at 1037a18–20, Aristotle *goes on*, at 1037a21–b7, to state firm if no doubt provisional conclusions on the topics under discussion in Z10–11 (how there is a *logos* of the composite whole). One may contrast the end of Z13, where after the new beginning signalled at 1038b1ff., Aristotle does point to the difficulties that stem from the denial of universal as substance for definition and does explicitly refer forward to a later discussion, 1039a14ff., 20ff.

arguments of those two chapters, as he does at 1037a21ff., he is quite firm. There he claims to have explained the way in which there is, and the way in which there is not, a *logos* of the composite whole, where it is clear that by *logos* he means definition, not merely account, and where his recommendation corresponds to the narrow definition according to which matter is excluded. 'With the matter', he says at 1037a27ff., 'there is no (*logos*) – for it is indeterminate: but according to the primary substance there is one, for example of man the *logos* of soul ... But in the composite substance, such as snub nose or Callias, matter too will be present.' To recapitulate Frede's point, Aristotle might be saying that no definition of a *sunolon* is possible (the only things you can define are forms) or he might be saying that the only way in which a definition of a *sunolon* is possible is by specifying its form or essence. But *either way* matter is excluded.

Full weight should be given to the fact that there is something of a discrepancy within the *Metaphysics* itself on the kind of definitions of animals and their parts that are to be given. Some texts treat them as a *tode en tōide* and point to the broader notion of definition, while others clearly recommend the narrower – not just in general but in relation to such items as animal and man in particular. It is obviously tempting to resolve this discrepancy by privileging one set of texts and treating it as canonical, the other as deviant, or otherwise by constructing arguments that Aristotle might have used to square his apparently divergent recommendations. However, to the latter point it has to be said that that would have to *be* a *construction*. Moreover, so far as privileging one group of texts goes, there is not just no clear indication as to which to take as canonical, but no indication whatsoever. In these circumstances we should rather contemplate the possibility of genuine hesitations on the subject on Aristotle's part.

However, since my principal task in this section is to review Aristotle's actual practice in the zoological treatises, the first issue that must be addressed is the extent to which they exhibit a concern with definition at all, and if so what the focus of that concern is. Pierre Pellegrin, for instance, has recently argued that the zoology is essentially a 'moriology', and that whatever may be true of Aristotle's analysis of the differentiae of parts, he should not be seen as intent on giving definitions of animal species. Good examples of such definitions – Pellegrin points out with some justice[54] – are not to be found in the zoological works. Without entering into detailed debate on yet another controversial issue,[55] three remarks may be in order. First, in Aristotle's apologia for zoology in *PA* I 5, even if it is expressly said that the student of nature is chiefly interested in form rather than in matter, it is also stressed that he should study the whole nature (*PA* 645a30ff.). Secondly, the attack on the dichotomists in *PA* I 2–4 is an attack on their incorrect procedures in obtaining their account of animals (*PA* 642b30ff., 643a7ff., 17, b10ff., 644a10f.); they are criticised for mistakes in obtaining the animal kinds (a point conceded by Pellegrin), and that it would be strange if Aristotle were not concerned to do better in the same regard. Thirdly, the absence of good complete examples of definitions of animal species might be as much a sign of his realising the difficulty of the task as of his not wanting to undertake it.[56]

Those difficulties take us to the heart of the matter. He has plenty of criticisms to offer of the dichotomists' incorrect procedures.[57] By implication, certain points about the procedures he would appear to approve emerge, though problems certainly arise concerning how he sought to apply those procedures in practice.

It is clear, for instance, that an animal kind is not to be defined in terms of a genus plus a single differentia arrived at by a single line of division: rather it should be defined by a conjunction of a plurality of differentiae (e.g. *PA* 643b12). Again certain natural kinds, such as birds and fish, picked out in ordinary Greek, should not be split up, that is they should not be put into different divisions (*PA* 642b10ff.). Again division should be by items in the *ousia* and not by essential or *per se* accidents (*sumbebēkota kath' hauto*, *PA* 643a27f.);[58] for this he offers a geometrical illustration: one should

---

[54] See Pellegrin 1982, 1985 and 1986. (esp. p. 99)
[55] Cf. further Lloyd 1996, ch. 16.
[56] Those difficulties did not, however, deter him from making a series of remarks concerning the being and essence of certain animal kinds, and indeed of offering certain definitional accounts, see further below, pp. 60–1. Cf. further Gotthelf 1985 on what he calls partial definitions (though it should be observed that it is only *complete* definitions that will meet *all* the requirements on indemonstrable primary premises in the *Posterior Analytics*).
[57] There is a particularly clear and incisive analysis of Aristotle's critique of the incorrect use of *diairesis* in Balme 1987a, 69ff., setting out the implied positive recommendations at 74ff.
[58] Cf. *PA* 645a36ff., *HA* 491a9f. See Kullmann 1974, pp. 63ff.

not define a triangle by saying that its internal angles sum to two right angles (though the application of this principle to zoological kinds is not as clear as it might be). Nevertheless all this shows, as Balme has put it,[59] that Aristotle's 'aim is not simply to make out and identify but to grasp the substantial being of the object'. Some of his remarks clearly indicate that every species of animal should figure somewhere in the eventual division and indeed only once (*PA* 642b31ff., 643a13ff.). I have always thought that that goes to show that he did indeed have in mind, as his goal, a comprehensive and exclusive system, not merely a description of non-exclusive groupings (though many modern commentators deny that he ever even sought a comprehensive classification of animals).[60]

The rules of correct division, stated in general terms, may be clear enough. The difficulties arise, as I remarked, in their application. Two interlocking questions that prove remarkably recalcitrant are:

(i) Does Aristotle operate with, or presuppose, what I have called the narrow, or the broad, view of definition?
(ii) Granted the equation 'soul is to body as form is to matter', how, in practice, in defining (whether of the broad or the narrow variety) is that to be cashed out?

Here the application of some of the results of the first part of this paper raises some far-reaching questions that can be brought to bear on current controversies on the viability or credibility of Aristotle's philosophy of mind in general.

On the first question it might seem that the answer must be obvious. It must be the broad view, it might be thought, since if the narrow is adopted (where matter is excluded) there is a very obvious objection. If definition is in terms of soul alone, this would seem to provide a totally inadequate means of differentiating animals. Of course the differences between plants, (other) animals, and humans can be and are grasped through the faculties of soul (as when he says that 'we define animal by the possession of per-

---

[59] Balme 1987a, 75.
[60] See e.g. Lennox 1987a, 1987b Pellegrin 1982 and 1986 – though contrast Pellegrin's rejection of a taxonomic interest in Aristotle with his chapter on Aristotle's classification of animals. In Balme (1961, 212 = 1975, 192) allowance is made for the possibility of an eventual classification of animals, though in the revised version of the paper, 1987a, all reference to such an idea is omitted.

ception', *PA* 653b22ff.), but this will not be enough to secure the differences *within* animal kinds that Aristotle needs. After all, all animals possess the faculties of nutrition/generation and the sense of touch, and most also have all the other senses and the faculty of locomotion, even though only one has reason in addition. Rather it is, of course, at the very least the *modalities* of those faculties in the different kinds of animals that will enable us to differentiate them. Those modalities will include, for instance, their methods of reproduction (viviparous, ovoviviparous, oviparous, larviparous, etc.), their modes of locomotion (walkers, flyers, swimmers, corresponding to their organs of locomotion and whether they are biped, quadruped, footless, polypod) and their modes and organs of nutrition.[61] Those types of differentiae evidently do receive much attention and at *GA* 732b15ff. he is particularly concerned to point out the lack of correlation between organs of locomotion and modes of reproduction, insisting that the latter cannot be held to depend on the former.

So if we take into account, as we surely should, the diverse modalities of the faculties of soul, that provides a far less impoverished basis for differentiating the chief groups of animals that Aristotle recognised.[62] However, a further difficulty for the narrow view of definition is presented by his apparent endorsement of the point that the major groups identified by common usage[63] have been marked out, in general, by the *shapes* of their parts and of their whole body, while within each group the differences are ones of degree, of the more and the less, for example of bigger–smaller, softer–harder, smoother–rougher.[64] Here too, however,

---

[61] See e.g. the differentiation by modes of locomotion at *PA* 639b1ff. and at *MA* 698a5ff., and reference to the modes and organs of nutrition in the account of bloodless kinds in *PA* 678a26ff.

[62] When in the *Politics* Aristotle compares classifying constitutions with classifying animals, he envisages a procedure that first identifies the essential parts of an animal and then considers all the possible permutations of these, *Pol.* 1290b25ff.: here the essential parts are specified as the sense-organs, the parts responsible for receiving and elaborating the nourishment, and the organs of locomotion, a specification that keeps close to his usual view of the principal faculties of the soul.

[63] However, he recognised that some important groups had *not* been named before him, specifying in particular that this was the case with the major division between the blooded and the bloodless, *PA* 642b15.

[64] See Lennox 1987*b*. I would agree with Balme 1987*a*, p. 79, however, that the reference to morphology in *PA* I, e.g. 644b7ff. (cf. *HA* 491a14ff.) is to be taken to relate to popular criteria, not to Aristotle's own. At *Metaph.* 1044a9ff. he specifies that differences in the 'more and the less' relate to differences in the composite whole, not to differences in the

caution is needed. In practice differences in shape are cited often enough:[65] but we have to bear in mind that Aristotle's explicit doctrine is that the parts are in general for the sake of their functions. Every organ or tool, and every part of the body, is for the sake of something, and *that* is specified by some activity (*praxis tis*): and so the whole body is for the sake of some complex activity[66] and so in a way the whole body is for the sake of soul and the parts of the body for the sake of the function (*ergon*) towards which each is naturally directed (*PA* 645b14ff.).[67] On this basis it would appear that while, naturally enough, morphology is often used to distinguish animal kinds, it is primarily the *activities* of the parts that have the shape they have that interests Aristotle – a viewpoint that is compatible with the narrow as well as with the broad sense of definition. That shape by itself is *not* a sufficient defining characteristic is clear from what is said at *PA* 641a18ff., where he points out that once the soul has left the animal, none of the parts remains the same *except in shape.*

The best way to test how much, in practice, Aristotle was willing to include in animal definitions is to review the limited number of texts that use the vocabulary of definition, essence, and substance; though since, notoriously, *ousia* is said in many ways, and *logos* too, these texts provide evidence that has to be used with the greatest circumspection.[68] One such group of passages does not take us very far, although it does confirm that Aristotle uses the usual, *de Anima*, definitions of plant, animal, and human based on the faculties of soul they possess.[69] Again, when he says at *PA* 693b13 that

'substance according to the form': this would suggest that it is differences in the *sunola*, rather than in the forms or essences as such, that are in question when he gives an account of the different kinds of beak that birds have or the different lengths of their legs, while such differences have to be related to, and may be explained in terms of, differences in activities, life-styles, and feeding-habits, see e.g. *PA* 692b22ff., 693a10ff., 694a1ff., b12ff., *IA* 714a21ff.

[65] See e.g. *PA* 692b8ff.
[66] Reading *polumerous* with Peck at *PA* 645b17.
[67] Cf. *PA* 641a1ff., 642a9ff., 646b10ff., 655b20ff., 687a7ff. and especially 694b13ff. (nature makes instruments for the work and not vice versa).
[68] There is a partial collection of texts from *PA* II–IV and *IA* in Gotthelf 1985, a survey from which he has, as he says, p. 28, omitted some 'methodological or otherwise theoretical' statements.
[69] At *PA* 653b22ff., for instance, he says 'we define animal by the possession of perception, and first of all by the primary one, that is touch', a point repeatedly made elsewhere, e.g. *Sens.* 436b10ff., *Somn.* 454b24f., *Juv.* 467b24f., 469a18ff., b3ff., *PA* 666a34, *GA* 731b4f., 736a30f., 741a9f., 778b32ff. Cf. *PA* 686a27ff. on humans and *GA* 731a24ff. on plants (with *GA* 715b16ff., where the *ousia* of testacea is said to resemble that of plants).

flying belongs to the *ousia* of a bird, and at *PA* 695b18f. that the nature of fish, according to the *logos* of their *ousia*, is to be swimmers, this too fits the narrow view of definition well enough, where it is limited to the faculties of soul, in that both these cases refer to the modalities of the faculty of locomotion.

However, a test case for how much might be included in definition is provided by references to such factors as *being blooded* or *bloodless*, mentioned in both the bird (*PA* 693b6) and fish (695b20) examples, and said more generally to belong to the account that marks out (*horizonti*) the *ousia* of the blooded and the bloodless animals at *PA* 678a33ff. How are these texts to be interpreted? One view takes it first that when being blooded or bloodless is said to belong to the 'account that marks out the being' of certain animals, it is indeed a *definition*, not just some weaker sense of 'account' that Aristotle has in mind. But given that the blood is said to be the matter of the whole body,[70] the conclusion we should draw, on this line of interpretation, is that we have strong evidence here for the broad sense of definition,[71] with Aristotle appearing to insist that a definition of an animal kind should include reference to the matter as well as to the form. The *ousia* thus marked out is the composite whole, *sunolon*, and – despite the recommendation of *Metaph.* Z 11, 1037a26ff. – it is to be defined by both components, not by the form alone.

However, that line of interpretation may be less secure than appears at first sight, and not just because of the conflict with *Metaph* Z 11. Passing over some minor difficulties,[72] I turn to one that arises from consideration of a text in the criticism of the dichotomists, *PA* 643a1ff. There Aristotle puts it that it is not possible for a single indivisible *eidos* of being to belong to animals

---

[70] See above n. 12.

[71] In the sense 'define' the verb *horizesthai* is usually used in the middle, not in the active, as at *PA* 678a34 in our passage, where the sense may be looser, 'mark out'. However, even in the middle the verb is used of different kinds of definition, notably at *de An.* 403a29ff. where Aristotle first describes and then criticises the definitions of 'anger' that on the one hand the *phusikos*, and on the other the *dialektikos*, would give.

[72] One such is that in the discussion of fish, Aristotle first says that their nature is to be swimmers according to the *logos* of their *ousia*, where *ousia* can be taken as essence or primary substance in the *Metaph.* Z sense: but then two lines later he remarks that fish are 'blooded according to their *ousia*', where, on the line of interpretation suggested, this should be glossed as the *sunolon*, *PA* 695b17ff. (cf. *PA* 693b6ff., 13 on birds). But if this is a somewhat harsh transition, one may note that it is no more so than many in the *Metaphysics* itself.

that differ *eidei*: rather it (the supposed indivisible *eidos* of being) will always admit of differentiation. His particular concern is to undermine the use of privative terms as differentiae, and quite how he can rescue his own use of such terms (e.g. bloodless) is itself controversial.[73] However, his argument proceeds by taking two positive examples, bipedality and bloodedness. Bipedality in a bird and bipedality in a human are to be differentiated (643a3f.), and there is no doubt at all that he does indeed endorse this differentiation; when he comes to consider the question at *PA* 693b2ff., for instance, he distinguishes between the inward-bending two-leggedness of birds and the outward-bending two-leggedness of humans. But his remark about bloodedness is that if they (the kinds that are blooded) are blooded, the blood is different, or else the blood is not to be included in the *ousia* (*PA* 643a4f.) – where we should observe that he says not that bloodedness is a genus represented in different species, but that the blood itself is different.

Of the two options mentioned, I think it is clear that Aristotle himself takes the first: the blood is indeed to be differentiated, and as we have already seen in Section 1, he offers a careful and detailed account of the differences in the quality of the blood of different-blooded animals and insists on the importance of doing so since the blood is responsible for many characteristics both of the character and of the perception of animals.[74] Moreover, if that is the option he favours, the one he *rejects* (on this interpretation) is that blood is no part of the *ousia* (*PA* 643a4f.) – where he can hardly be rejecting the idea that blood is part of the composite whole and presumably has in mind that it is no part of the essence.

What this might be taken to suggest is that although blood is said to be matter, it has a role in the being of certain animals that that by itself does not capture. Rather, bloodedness has to be differentiated and the differentiations that will mark out the blood of one species from that of another contribute to the activities of each[75] – and *mutatis mutandis* the same will apply also to what is analogous to blood in the bloodless groups. But if that might suggest a line of argument that would go to save the principle sug-

---

[73] See e.g. Balme 1972, pp. 110, 120f., Gotthelf 1985, pp. 34f.

[74] See above, pp. 44f. and n. 16, and cf. e.g. *PA* 679a25f. on bloodless animals.

[75] At *PA* 647b29ff. Aristotle remarks generally that there are differences in the uniform parts that are 'for the sake of what is better', exemplifying this with the variations in the blood: cf. *PA* 648a13ff.

gested in *Metaph*. Z 11, that definition is of the *form*, it would do so only at the cost of highlighting the problem that may be said to emerge as one of the by-products of our analysis in Section 1. If *bloodedness* belongs to the being as essence (and not just to the *sunolon* as the composite of form and matter) how far does the essence or the form extend? What I have called the narrow view of definition threatens to become indistinguishable from the broad: or in other words the threat is to the very application of the form–matter dichotomy in the case of living creatures.

There is no doubt that the form–matter dichotomy is needed in the zoology, and not just for any definition of an animal kind or animal part that Aristotle might wish to attempt; nor just for the sake of the standard analysis of substances as *sunola* of form plus matter (and, after all, living creatures are the paradigmatic and primary examples of substances in the *Metaphysics*). Aristotle also needs that dichotomy in zoology in, for example, his theory of reproduction, where the form is contributed by the male parent, the matter by the female.[76] However, the problem is clear from a consideration of the differences in the way in which the dichotomy applies in the two main spheres to which it is applied, to living creatures on the one hand, artefacts on the other – difficulties that Ackrill, among others, has remarked.[77] Of course, even in the case of artefacts, the matter of which a house is built is not just matter, certainly not 'prime matter' (if, as is controversial, that corresponds to an idea that can be reached by abstraction; though all are agreed that it is nowhere encountered in the world). Rather, the matter is bricks or stone or wood, each of which has certain characteristics that differentiate it from other kinds of thing. But in the case of artefacts the bricks or stone have the characteristics they have (as bricks or stones) whether or not they are incorporated in a house.

But that is not true of the material parts of living creatures, neither the uniform parts, flesh, bones, blood, and so on, nor the non-uniform ones. As Aristotle says in so many words at *PA* 645a35ff., the material the biologist has to deal with are things 'that do not even occur separated from the being itself'. *All* the

---

[76] See above, p. 42 and n. 7 and cf. Code 1987.
[77] Ackrill 1972–3/1979. Cf., however, e.g. Williams 1986, Lear 1988, ch. 4.
[78] See above, n. 5.

parts of the body have soul:[78] all (with the exception of the useless residues) contribute in one way or another to the vital faculties of the living creature. In practice, in zoology, as he says at *PA* 643a25, no part is *just* matter. Unlike what is true of the analysis of artefacts, the zoologist has to study composite wholes whose constituent *material* parts cannot be said to have the characteristics they do as the material parts they are *outside* the composite whole: their characteristics cannot be properly specified independently of the form, that is the vital faculties.

We are used to the notion of the correlativity of form and matter; what will count as form, what as matter, will depend on the *whole* under consideration. But in the case of living creatures, the fact that their (proximate) matter is not independently identifiable as the matter it is outside living creatures (dead flesh is no flesh) introduces a fundamental difference. For Aristotle to talk as he does of a man being decomposed into bones, tendons, and flesh (e.g. *Metaph.* 1035a18ff.) is from one point of view to talk loosely: for the man is not destroyed into flesh and bones but into what is only homonymously flesh and bones (and even if he had specified blood as the matter, the same point would apply).

From the vantage-point of the issues we have considered, some final remarks may be ventured on the current debate between functionalists and antifunctionalists on the question of Aristotle's philosophy of mind and its credibility. First, a series of texts demonstrate, I should have thought conclusively, that the antifunctionalists must be right on the basic point: there can be no question of the souls/forms of living creatures being realisable in matter other than the matter in which they are found, and what has been called the 'compositional plasticity' of *psuchē* is minimal, if not zero.[79] On the other hand, that is certainly not ruled out, indeed it is emphasised, in the case of artefacts – within limits of course: the same form of a house can be realised in either bricks or stone or cement (even if not in pure sand or in water: the saw, as Aristotle put it, not in wood nor wool[80]), and that shows that the expectations that the functionalists entertain are expectations that have

---

[79] When the possibility of the eye being constituted by air (rather than, as he believes, water) is mentioned at *Sens.* 438a13ff., b3ff., this is not in the spirit of investigating 'plasticity' but part of a complex argument from exhaustion to show that it *must* be of water. Contrast Code 1991.

[80] *Metaph.* 1044a29.

a genuine Aristotelian basis, even if not one in his psychology or biology.

Burnyeat argues that since the Aristotelian theory of matter has been exploded by science since Descartes, we cannot consider his theory of mind credible. But that seems both too hard and too soft. Too hard, maybe, because it is not everything about Aristotle's theory of matter that is at fault. (One might add that as between Aristotle and Descartes's own view of matter, there may not be too much to choose between them from the point of view of *modern physics*, whatever we say about modern philosophy of mind: a philosophy of mind that aimed to take modern physics fully into account could not afford to start from a Cartesian mind–matter contrast any more than from an Aristotelian one.) But too soft, because even for Aristotle there is a problem. This is not the quasi-evolutionary problem of the emergence of life. But a problem exists, none the less, in reconciling the artefact model of *sunolon* with the vitalist one, of reconciling one view that has it that matter is independently identifiable (in the sense that what counts as the matter has the properties it has independently of the whole of which it is the matter) and one that has it that it is not. Of course *sunola* come at different levels of complexity, and so correspondingly do forms and matters. But the difference between the artefact model and the vitalist one is not one simply of grades of complexity (after all, there is nothing to choose between houses and dogs, or between bricks and blood, on that score) but one of the nature of the analysis of matter to be given: is it, or is it not, dependent on the *sunolon* or the form? The problem is the more acute in that there is a perfectly well-unified theory of matter on offer from Aristotle's rivals, the atomists. Indeed, Aristotle himself up to a point offers a unified theory of compounds, the homoiomerous substances that consist of compounds of the simple bodies. They have all undergone *alloiōsis*, qualitative change, and this allows him to claim that, as the homoiomerous substances they are, they acquire new properties. Yet what that theory still fails to provide for are the differences we have remarked between flesh and bones on the one hand, and the inanimate homoiomerous substances such as gold on the other.

Whether or not we find Aristotle's theory of mind credible, the unity of his theory of matter, both of his theory of the elements and that of the homoiomerous parts, is under extreme strain in

Aristotle himself. The difficulties we have identified lead back to the issues we raised in relation to the ultimate constituents of material objects: the problem of reconciling what is true of earth, water, air, and fire *haplōs*, in an unqualified sense, and what is true of them as the constituent elements of living uniform parts: or again of reconciling what is true generally of hot, cold, wet, and dry, and of hot, cold, wet, and dry as what is 'responsible practically for death and life, and again for sleep and waking, for maturity and old age, and disease and health'. It is all very well to say that Aristotle has (as indeed he has) a top-down theory of life, but the problem when he *gets down* to the bottom level, to the general theory of material elements, is severe. He cannot afford two theories of the material elements, one for living things, the other for the inert. But to resist that conclusion he needs to explicate the relationship between the two, the relationship between, for example, vital heat, and heat, or between pneuma and air, or between pneuma in oil and pneuma in semen.[81]

These are problems, indeed, in his theory of matter. Burnyeat's paper focuses most usefully on that as the locus of important difficulties. But they were not problems that needed Descartes to discover, for some emerge already from a comparison with ancient atomism. Indeed, the hesitation in many of Aristotle's own pronouncements on just these issues can be taken to suggest that he already had a sense of the difficulties. To my mind the hesitations or the vacillations that we detect point to that conclusion, to a realisation on Aristotle's own part of some of the difficulties the complex of theories we have been discussing presented. I have focused in particular on the problem that arises in attempting to resolve the question of what is to be included in the definitions of living creatures – where both narrower and broader views are to be found both inside and outside the zoological treatises. Again I remarked on the problem of demarcating the living from the non-living. Any erosion of the firm distinctions between non-living and living, and between plants and animals (distinctions that correspond to the clearly demarcated differences in the faculties of soul that they possess), would be a major source of embarrassment for Aristotle's theory: and yet in practice he clearly allows that

---

[81] See the texts cited above, nn. 22–3. For *pneuma* in semen, see e.g. *GA* 735b33, 736a1, 9, cf. 737b26ff.: for *pneuma* in various inanimate substances, see e.g. *Mete.* 383b20ff., *GA* 735b19ff., 761b11, 762a18ff. (the last two passages in connection with spontaneous generation).

there are problematic cases.[82] Again, although he says less on the problem of the nature of vital heat, quite how far he has resolved to his own satisfaction its role in spontaneous generation, for instance, is not clear.*

It is undeniably one of Aristotle's great originalities to have introduced matter as the correlative to form. But to win his point against Plato (as well as against the atomists) he has both to distinguish clearly between form and matter and to insist on the unity constituted by the two, the unity of the *sunolon*. Of his two main models, the artefact model does an excellent job illustrating the *distinction* (especially where the matter can be identified by the properties it has independently of the *sunolon* in which it figures). Conversely, the *unity* of form and matter is seen most clearly in the case where the form is the soul, the matter the living body of which it is the actuality.[83] Yet the greater the confidence Aristotle expresses in the unity, the greater the tensions that arise in the application, in practice, of the very idea of the distinction between form and matter. The more the vitalist model dominates, the more hylozoism threatens. But the more the contrast between the vitalist and the artefact model emerges, the more difficult it is to dismiss the competition provided by atomism. That he still works with the form/matter distinction in the zoology is abundantly clear: that there are difficulties in his so doing I have tried to show: that some of these are such that he himself was aware of them (in a way quite close to the form in which I have presented them) is I think likely.[84]

Postscript

The issues of functionalism and anti-functionalism have been the subject of considerable debate since this article was written, not

---

[82] Cf. the texts cited above, n. 25.**

[83] The problem of the unity of form and matter in the composite whole is related to, but importantly distinct from, the problem of the unity of genus and differentia discussed at *Metaph.* Z 12, 1037b10ff., H 6, 1045a7ff. It is striking that to resolve the latter he appeals at 1045a23ff. to the relationship between form and matter, actuality and potentiality, *as if* their constituting a unity were less problematic.

[84] An earlier draft of this chapter was read to the Southern Association for Ancient Philosophy meeting at Cambridge in September 1988. I would like to thank all those who participated in the discussion, and Myles Burnyeat, Malcolm Schofield, and Robert Wardy for subsequent further detailed comments.

* This problem will be discussed in some detail below, ch. 5.

** Cf. below, ch. 3.

least from other contributors to the collection in which it was originally published: see especially Nussbaum and Putnam 1992, Whiting 1992, with which cf. Charles 1988. Whiting, in particular, has suggested a distinction between two senses in which the matter of a living animal may be spoken of, (1) as the organic body that is essentially ensouled, and (2) as the quantity of elements accidentally ensouled. The former corresponds to the sense in which the anti-functionalist would insist that the relationship between form and matter in the living creature is not contingent, while the latter may suit a functionalist view that it is.

It has also been objected to me (David Charles, personal communication) that my dismissal of functionalism was too swift in one respect. If, contrary to the assumption I made above, p. 62, all that functionalism requires is that the possibility of alternative realisation of form in matter be an epistemic, not a metaphysical, one, then Aristotle's readiness to consider such a case as an eye constituted by air at *Sens.* 438a13ff., mentioned in note 79, can be given more positive weight than I allowed. That alternative realisation is indeed ruled out in fact, but it is at least contemplated.

Further issues to do with the possible threat of a collapse of Aristotle's element theory into hylozoism will be addressed in later chapters (4 and 5) as also will other aspects of the ongoing controversies concerning the interpretation of Aristotle's account of perception (chapter 6).

CHAPTER 3

Fuzzy natures?

We know how Aristotle generally seeks to define humans, animals, plants, that is by their vital faculties. Humans possess reason, animals (including humans) perception, plants the nutritive/reproductive faculty, *threptikē*, alone. All that is clear: or is it? At points there are hesitations in Aristotle's remarks, and I want to investigate why, focusing especially on two famous texts in the *Historia Animalium*, 588b4ff., and the *de Partibus Animalium*, 681a9ff., in which he raises problems directly about the borderlines between the non-living and the living and between plants and animals. But before plunging into the analysis of those texts, we should first distinguish different types or grounds for hesitation.

First, hesitation may reflect (merely) our not knowing. There is some obstacle to our knowing, the difficulty of carrying out certain observations, for example. Perhaps not all the evidence is yet in for us to decide: it may be that it is in principle impossible to get at all the data. But that does not mean that there is any doubt that there is a determinate fact of the matter there to be known. We may call this epistemic hesitation.

Contrast that with what I shall call ontological indeterminacy. It is not just that we cannot determine hard and fast boundaries for lack of crucial data: the boundaries themselves are indeterminate or there may be no boundaries, as such, at all. The penumbra round a shadow, or the edges of a cloud would illustrate the former. The sounds produced on a single unfretted string, when a bridge is moved continuously along its length, would illustrate the latter (though at any given point the sound will have a certain determinate pitch).

If we apply that contrast to the plants/animals case, we can see a radical distinction between two opposing lines of interpretation. On epistemic hesitation, there are plants, and there are animals, and even though there are certain living beings where we are at a loss to say to which category they belong, they must belong to one or other. We may not be able to tell, for instance, whether they have the sense of touch. But if we could answer that question, that would give us an unequivocal answer to whether they are plants or animals.

On ontological indeterminacy, by contrast, there could be beings that are intermediate between plants and animals, neither the one nor the other, or at least neither fully the one, nor fully the other, and not just because we cannot tell, but because they are like that. Such cases in turn may or may not mean that the division of all terrestrial living beings into animals or plants has to be revised. The classification might be saved, at a price, by treating it as normative: exceptions should not be allowed to count against it, as they are just deformities. Or else the two classes may not be the hard-edged, mutually exclusive and exhaustive categories that is generally assumed. What it is to be a plant or an animal may be a matter not of a single decisive criterion, but of fulfilling or not fulfilling several. If there are indeed complex and multiple criteria both for animality and for planthood, there may be items that are intermediate in the sense that they meet some, but not all, of either class. The classes would have a penumbra, then: they would be less determinate, more cloud-like or fuzzy than is usually allowed for.

The texts we shall be considering certainly express epistemic hesitation: that is not in question. But they also appear to suggest ontological indeterminacy. The issue is, do they, and if so, what are the implications, not just for his zoology, but for his methodology and philosophy of science?

Given the confidence that Aristotle generally exudes with regard to the contrast between plants and animals and his repeated appeal to the criterion of perception, one might suppose that there could be little to say in favour of the second line of interpretation in this case (at least) and that any hesitation Aristotle expresses is merely epistemic. So it is first worth reminding ourselves that there are modes of ontological indeterminacy of an uncontroversial kind in Aristotle.

Matter, *hulē*, to start with, is indeterminate, *aoristos*. According to a well-known text in *Metaphysics* Z 10, 1037a27ff., discussed before (chapter 2, p. 54), there is no definition (indeed no *logos*) of the *ousia* taken as the *sunolon*, composite whole, in one sense, even though there is in another. There is definition of the essence, of what is called in these chapters of *Metaphysics* Z the *prōtē ousia*, primary substance, of the *eidos to enon*, the inherent form, and the example given is that of human, where there is a *logos* of the soul. But there is no definition of the composite whole, the *sunolon*, taken with the matter, since, precisely, it – the matter – is indeterminate, *aoriston*. We have noted before, following Michael Frede,[1] that this restriction can be taken in two different ways. *Either* Aristotle could be saying that there are no definitions of composite wholes *since* they contain matter. *Or* he could be saying that there are no definitions of them *insofar* as they contain matter: that is, that the material part of the composite is not definable, even though there are definitions of composites insofar as they contain forms. But *either way*, matter, on this story, is indeterminate and indefinable.

There is something of a tension between that text in *Metaphysics* Z and the equally well-known discussion of *phusikē* and *phusika* in *Metaphysics* E 1 (also mentioned above, chapter 2 at n. 49), where it is said of some things that are defined, and some essences, that they are like snub, others that they are like curved (1025b30ff.). The latter have no perceptible matter, the former do, and indeed every kind of natural object (*panta ta phusika*) is said to be like snub, i.e. with perceptible matter. That gives him the contrast he is interested in there between *phusikē* and *mathēmatikē*, but while allowing definition of composites of *some* type does not address the difficulty later raised in *Metaphysics* Z 10. However, the problem of reconciling the two accounts need not concern us here, since the main point for our purposes is clear. For Aristotle matter or the material is what is *in*formed, and as such, as the matter it is, is indeterminate. To use the language of Plato's *Philebus*, it is *apeiron*, the as yet to be determined, while the form is the factor that does the determining.

Here then is a major manifestation of the indeterminate. But *by the same token*, the form is as determinate as anyone could wish. Insofar as we are dealing with forms, then, in physics just as much as in mathematics, there should be no ontological indeterminacy about them, and any indeterminacy in Aristotle's statements in their regard will have to be put down to epistemic hesitation.

---

[1] Frede 1985; cf. Frede and Patzig 1988.

Yet that does not do justice to several complexities. Form and matter, as we all know, are correlative to one another. But Aristotle accepts, and insists on, differences between higher grade and lower grade forms. You do not get matter on its own, of course: but some items have very low-grade forms.[2] The argument about the unity of proper substances in *Metaphysics* Z is an argument that protests that mere *heaps* will not count as such (1040b8ff.). Not until they have been concocted, as he says, in a remarkable use of the notion of concoction (see further below, chapter 4). Only entities with a well-marked unity, provided by a determinate form, will count.

In practice, in his extensive discussion of the homoeomeries in *Meteorologica* IV, Aristotle often talks very loosely of those kinds of material objects in which water predominates, those in which earth does, and so on. The former are spoken of as 'kinds of water' (e.g. *Mete.* 382b13, exemplified by wine, urine, whey), or they are said to be 'of' water, or to belong to it, using the genitive, *hudatos* (e.g. *Mete.* 383a6f.). Other things are said to be common to earth and water (*koina gēs kai hudatos*, e.g. *Mete.* 383a13f.) where they may of course have more of one or more of the other (*Mete.* 383b18ff.).

Of course Aristotle's 'simple' bodies are not chemically pure substances and their compounds are not determined by chemical formulae of the H_2O type. He discusses which types of materials may be dissolved, split, solidified, thickened, softened and so on. These differentiae yield a very rough classification, or at least some broad groupings. Thus most metals are 'of' water, but then so too are glass and many – 'nameless' – stones that are melted by heat, *Mete.* 389a7ff. But iron, by contrast, is rather, or preponderantly (*mallon*), 'of' earth, as also, however, are horn, bone, leaves, amber, soda, salt and non-meltable stones, *Mete.* 389a11–19. With the homoeomeries we are evidently dealing with comparatively low-grade forms. Even though some of them have been 'concocted', nevertheless, according to the criteria of *Metaphysics* Z 17, 1040b5ff., they will not rate as proper substances, for by the strict requirements on unity he there demands, he rejects not just the simple bodies, but also the parts of animals, as such.

Further elements of indeterminacy figure in other parts of Aris-

---

[2] How far down do forms go? That even the simple bodies have form may be suggested by *Cael.* 276b21ff., 277a1ff., 8ff., where *eidos* evidently does not mean shape. Yet they are not (proper) substances according to the argument of *Metaph.* 1040b5ff., which also rules out the parts of animals.

totle's account of qualitative differentiae. His theory of perception, for instance, is, to be sure, a theory of the reception, by the sense-organ, of the perceptible form without the matter of the object perceived. But in his account of perceptibles, in the *de Sensu*, he first seeks to associate each of the sense-organs with one of the simple bodies. In ch. 2, 438b19ff. his conclusion is that the sense of sight is 'of' water, hearing 'of' air, smell 'of' fire and touch – and taste – 'of' earth, on each occasion using the genitive, no doubt to suggest a looser connection than a straightforward identification of the sense-organ with the simple body. But then on the side of the perceptibles themselves, he treats colours, for instance, as total mixtures of black and white (440a31ff.). Exploiting an analogy with concordant sounds, he suggests that the pleasant colours correspond to determinate ratios (such as 3 to 2, or 4 to 3) (439b25ff.). But while that shows his ambition to divide colours up into determinate types,[3] it leaves plenty of indeterminacy in the actual colours we experience, insofar as many of them will fall outside those provided for by the simple ratios.

As one goes down the scale of being, the elements of determinacy fade, as it were, and one deals with what is less determinate, has less form, is more 'material', even though never just matter, let alone pure matter. But our primary concern here is with plants and animals, and they are surely determinate enough, indeed the prime examples of proper substances fulfilling the strictest unity requirements of *Metaphysics* Z 17.

Indeed that is so. But the first point worth making concerns how the 'more and the less' are used in relation to these substances. I do not just mean that there may be accidental differences between individuals of the same species, as, for example, when one human being is tall, another short. The 'more and the less' are introduced more technically as a mode of differentiation between species belonging to the same genus in *de Partibus Animalium* I 2–4. In his attack on dichotomy there he places within one *genos* groups that differ among one another only 'by excess' or 'by the more and the less'. Thus at I 4, 644a19ff., one kind of bird differs from another by being long-winged or feathered or short-feathered. But birds' feathers and fishes' scales differ more widely and correspond only 'by analogy' (cf. chapter 7).

[3] See Sorabji 1972; Barker 1981.

So differences in colour, shape, size and so on are allowed as marks that help to differentiate between kinds of animals within a single *genos*.[4] Yet those differences by 'the more and the less' are not the *chief* criteria invoked in determining the essences of living creatures, which are, rather, a matter of their activities, *energeiai*. Morphological differences are acknowledged, for example, at *PA* 644b7ff., *HA* 491a14ff., but they take second place to differences in the vital faculties of the creatures concerned (cf. above chapter 2 at n. 64). It is walking or flying that pick out what belongs to the essence, not doing so with long or short legs, long-feathered or short-feathered wings.

Here, surely, when dealing with such an item as the capacity to locomote in one way or another, to perceive with one or more of the varieties of perception, there can be no question of ontological indeterminacy. The application of the law of non-contradiction rules out both asserting and denying simultaneously the same determinate predicate of the same subject, and the law of excluded middle stipulates that we must either assert or deny it. That is ungainsayable. But the question for us is whether the types of difficulty we encounter in the zoology can be resolved merely by the application of those principles.[5] How far indeed do they apply? Are the statements we are concerned with well-formed formulae to which they *have* to apply?

Moreover even when we can – when we have all the evidence – expect a yes or no answer to whether a particular animal or plant has a particular vital faculty, that does not necessarily mean the end to the type of problems we have to deal with. Animal kinds, we are told in another passage in the criticism of the dichotomists, *PA* 643b1ff., are not reached by a sequence of dichotomous divisions, but are marked out by many differentiae cutting across such divisions. The classification is, in other words, polythetic.

---

[4] But, as Pellegrin 1982 and 1986 and Lennox 1987a especially have insisted, the terms *genos* and *eidos* are both used by Aristotle of groups, kinds or forms at varying taxonomic levels. Cf. Lloyd 1991 at pp. 373f.

[5] Balme 1991 in a note to *HA* 588b4 was confident that they could. He took *epamphoterizein* to mean 'tend to both sides', but asserted that it is impossible to 'be on both sides, either actually or logically'. But that is to treat all zoological propositions on the question as well-formed formulae and to discount the possibility that in a polythetic classification some criteria may be met, others not, for inclusion in a kind. While some of Aristotle's dualisers only dualise in appearance, and we can say that in their nature or in reality they fall on one side of a clear boundary, that is not so in all cases, nor, as I shall argue in the present instance of the plant/animal divide, is the boundary always clear.

Fuzzy natures?

But if that is the case, that leaves open the possibility of problematic cases, where no simple answer can be given as to whether animal X belongs to group Y, since by some criteria it may, by others not, but to group Z. Of course in such a situation various types of response can be tried out. Are the criteria invoked themselves equivocal and in need of disambiguation – as is the case, for example, with *pezon* ('walking') and *enudron* ('aquatic') discussed at *HA* VIII 2, 589a10–590a18. Should either or both of the groups Y and Z be subdivided, or further groups intermediate between them be entertained?

The problems, in such cases, will often need to be resolved, insofar as they can be resolved, not by a straightforward appeal to the laws of non-contradiction or excluded middle, but by an evaluation of the empirical data. They will require the acumen not of Aristotle the logician, but of Aristotle the field-worker. Nor are we entitled to assume that Aristotle the naturalist will always be in a position to arrive at neat solutions, not least in the principal case we have to consider, where there is, we can agree, every reason to be hesitant in the face of the difficulty of determining the boundary between what is an animal and what is a plant.

Aristotle's two main discussions, *HA* 588b4–589a2, and *PA* 681a9–b12, share three general features. First there is the hesitancy of his language, second certain clear indications of epistemic hesitation, third some other cases where the difficulties he mentions appear to relate not just to our not knowing, but to nature itself. I shall review these general features before turning to analyse in detail what he says about each of the problematic living creatures or beings he identifies.

First as to the hesitancy of his language. He repeatedly introduces remarks with *phainetai*, what appears,[6] for example *HA* 588b9, b22, b24, or with *dokei*, what seems or is thought to be the case, for example *HA* 588b8, *PA* 681a14 or in the more guarded form of the conditional plus *an*, *doxeien an*, *PA* 681a27. Or again he refers not to what something is, but to what it is like, with *eoike* at *HA* 588b17, b21, with *homoiōs echei*, *PA* 681a16f., (cf. *HA* 588b26, *homoiōs*), with *paraplēsion*, *PA* 681b7, with *hōsper*, *HA* 588b10, b31, *PA* 681a20.

Then among the passages that register what I have been calling

---

[6] The term can also mean 'is evident' and may indeed do so at *HA* 588b24, see below p. 79.

epistemic hesitancy, that is the difficulties of our deciding, are: *HA* 588b5f., the boundary and the middle, between the non-living and animals, escapes our notice, *lanthanein*; *HA* 588b12f., as regards certain things in the sea, one would be at a loss, *diaporēseien an tis*, whether they are animals or plants; *HA* 588b17f., as regards perception, some of these give not a single indication of it (*sēmainetai*), others only faintly;[7] *HA* 588b26f., similarly in the case of some animals, one can apprehend no function other than generation (*ouden estin ... labein*); *PA* 681a28, how to classify them is unclear, *adēlon*; *PA* 681a31, the creature has no clear (*dēlon*) residue; *PA* 681b7f., in virtue of their having no apparent (*phaneron*) residue.

At the same time, at points there appears to be more than just epistemic doubt in play, when Aristotle refers to what nature does or what is the case. *HA* 588b4ff. opens with: 'thus nature moves from the non-living to animals little by little' and goes on to refer to the continuity of this transition (*tēi sunecheiai*, b5). A few lines later he says: 'the transition from them (he is here referring to plants, cf. b7) to the animals is continuous, *sunechēs estin*, as was said before' (*HA* 588b10–12). Similarly *PA* 681a12f. comments on the cases of the sea-squirts (*tēthua*) and sponges that he has just mentioned with: 'For nature moves continuously from lifeless things to the animals through things that are alive but not animals.' As for certain kinds of sea-anemones, *knidai* or *akalēphai*, he says at *PA* 681a35ff., that they are not testacea, but fall outside the groups reached by division, but they 'dualise' (*epamphoterizei*) between a plant and an animal *in their nature* (*tēn phusin*).

So while some passages in both discussions relate (just) to the difficulties of our knowing what is the case – difficulties that could be settled, one way or another, if more or better or clearer evidence were available – others appear to commit Aristotle to the notion that there is a continuity (*sunecheia*)[8] in nature itself, between the non-living and plants and again between plants and animals.

---

[7] Reading *oude hen*, rather than Balme's *ouden*, though the sense is not substantially altered.
[8] It may be noted that *sunecheia*, *sunechēs*, may be used (1) of contiguity, where two items are in contact (cf. *haptomenon*), with no gap between them (cf. Balme 1991 p. 61 n. a), or (2) of strict continuity, as of the continua of space, time or geometrical magnitudes, where in an infinitely divisible line, for instance, no two points are adjacent, and in a line divided at a point, each segment so formed shares the point as a limit. Clearly to understand how Aristotle uses the term in our context, our best recourse is to investigate the actual cases he mentions, where the upshot will be that there are intermediate species that meet some, but not all, of the criteria for classification as animals or plants.

How can that be? To answer that we have to examine the individual cases that cause the difficulties. But it is important to note first that there is no suggestion, in our two texts, that the problems could be resolved by postulating a third class, intermediate between plants and animals. Rather the inquiry, in both cases, focuses on the question as to which side of the plant/animal boundary a particular item comes, which class it falls into. *poteron*, he asks at *HA* 588b13, and again *poterōs* at *PA* 681a28, cf. *poterōn* at *HA* 588b6 with regard to the boundary between the lifeless and the living, first plants and then animals. Even though there are those 'dualisers' at *PA* 681a35ff., he would not need to speak of them as such if they belonged clearly to some third group, as opposed to – apparently – having a certain tendency to belong to both.

Six main types of living being occupy Aristotle's attention, where we can, of course, draw on what we are told elsewhere in the zoological works as well as on what is said in our two principal texts. These are (1) sponges (*spongoi*), (2) *tēthua* (sea-squirts or ascidians), (3) *pneumones* (or *pleumones*) – 'sea-lungs' (jellyfish or medusa) – with which we can group the otherwise unidentified *holothouria* with which they are associated, (4) the *knidai* or *akalēphai* (here, *PA* 681a36, treated as alternative names for the same living being) – sea-nettles or sea-anemones, (5) the pinna (*pinna*) and (6) razor-shells (*sōlēnes*).

(1) As to the sponges, first. The verdict of our two main texts is very similar. *HA* 588b20f. says that they completely resemble plants: *PA* 681a11f. that in every respect they have the function (*dunamis*) of a plant, and again a15ff. that they are completely like plants. That is not to say that they *are* plants, though the main reason for not identifying them as plants that is given elsewhere is not mentioned in our two texts.[9] This is that the sponge seems, or is thought (*dokei*) to have some perception. At *HA* 487b10f. the evidence (*sēmeion*) of this that is offered is that – as they say, *hōs phasin* – it is more difficult to tear them away unless the movement is done surreptitiously. At *HA* 548b10ff., after reporting that 'they say' that it has perception, citing the same reason, he goes on to note

---

[9] Balme (1991 note to *HA* 588b4) took it that references to the sponge's sensation, at 487b9 and 548b10, are an indication that those reports are later than *HA* 588b20f. and *PA* 681a15. But that does not seem necessary: though our two texts emphasise the resemblance between sponges and plants, neither asserts that sponges are plants.

that some dispute this, for instance those at Torone (whose view gets the support of D'Arcy Thompson who dismisses the story about its resistance as mere fable, 1947, p. 250). Yet a little later, 549a7ff., Aristotle says of one kind, the 'unwashable' (*aplusia*), that they are most especially agreed by all to have perception.

But against the sponges being animals should count the fact that they only live 'attached' (*prospephukōs*) and do not live when 'detached' (*apolutheis*), *PA* 681a15ff. *HA* 548b5ff. specifies that they all grow on rocks or sand-banks and get their nourishment in the mud. At *HA* 548b17f. Aristotle speaks of its 'roots' (*rizas*), and says that when torn off, it grows again from the portion that remains. It is evidently in virtue of this 'non-detachability', as I shall call it, that in *PA* 681a15ff. it is said to be 'completely like plants'.

(2) The sea-squirts 'differ a little in their nature from the plants' but are 'more animal-like (*zōtikōtera*) than the sponges' *PA* 681a9ff. They are usually classed with the testacea, one of the regular groups of bloodless animals that Aristotle recognises, e.g. *PA* 680a4ff., but they present certain problems, and not just those associated with the testacea as a whole, when they are said, for instance, to have their heads downwards, like plants (*PA* 683b18ff.). There is a doubt, first, about its perception. At *HA* 531a8ff., the fullest description, its general lack of instrumental parts and sense-organs is noted, a27ff., though it is said to have a fleshy part, *sark-ōdes*, a17f. That is mentioned again at *HA* 588b19f., just after Aristotle had referred to the weak, or non-existent, signs of perception in some kinds, and at *PA* 681a27f. it is in virtue of its having this fleshy part that Aristotle is prepared to hazard the inference that it has perception. It produces no apparent residue, however, *HA* 531a14f., a point repeated about it and the other testacea at *PA* 681a31ff., and there used to justify calling it, and any similar animal, plant-like, *phutikon*, since none of the plants has any residue. Moreover it too (like the sponge) only lives 'attached', viz. to rocks *HA* 531a11ff., cf. a19ff., and in this is said at *PA* 681a25ff. to be similar to a plant.

(3) The jelly-fish, which are only mentioned in the second of our two main texts, in *PA*, present a strikingly different set of characteristics. At *PA* 681a17ff. he contrasts the 'so-called holothouria' and the jelly-fish and any other such-like beings in the sea with the sponges which he has just been talking about. They differ a little from them (the sponges) in being 'detached'. But, he goes on, 'they

have no perception, but live like plants that are detached' (*hōsper onta phuta apolelumena*) (*PA* 681a19f.).[10]

(4) At *PA* 681a35ff. *knidai* and *akalēphai* are evidently alternative names given for the same creature, though elsewhere we have references to several kinds of both. At *HA* 548a24ff. there are two groups of *knidai*, one that do not detach themselves from the rocks, the other that do get detached from smooth flat ledges and that move about. In a corrupt passage, *HA* 531a31ff., b10ff.,[11] there is a distinction between two groups of *akalēphai*.

In the first of our two main passages, *HA* 588b19f., the *akalēphai* are mentioned, along with the sea-squirts, as having a fleshy part, just after Aristotle has talked of the lack of indication, in some kinds, of perception. Then at *PA* 681a35ff., where the *knidai/akalēphai* are said to 'dualise' in their nature (see above, p. 74), we are then told why. On the one hand, 681b2ff., some are detached and fasten themselves on their food, they are animal-like (*zōiikon*), and also in that they perceive what comes up against them (cf. *HA* 531b1f. on the *akalēphai*) and furthermore they use the roughness of their body as a means of self-preservation (cf. *PA* 683b9ff. on other testacea as well). On the other hand, 681b5ff., in virtue of their being incomplete (*ateles*) and quickly attaching themselves to rocks, they are like the class of plants, and he adds as also in having no residue (even though they have a mouth): those last points also appear elsewhere, as at *HA* 531b8ff. certain *akalēphai* appear to have no residue at all and in this respect, *kata touto*, are like plants, while at *HA* 590a28ff. they are said not just to have a mouth in the middle (which is clear in the bigger ones) but also, like the oysters, a vent by which the nourishment is excreted.

(5) As for the pinna, it, like the sea-squirts, is normally treated as one of the testacea, e.g. in *HA* v 15, 547b15ff., where, following Thompson's text, they are said to grow out of their *bussos* in sandy

---

[10] As with the remark about sponges discussed in the previous note, to say that the *holothouria* and jellyfish live like plants that are detached is not to identify them as plants. On the contrary, so far as the *holothouria* go, in the only other reference to them, at *HA* 487b15, Aristotle clearly treats them as animals. They, and certain kinds of oysters, are cited as among the stationary (*monima*) animals *zōia*, 487b6: they are detached but 'immobile' (*akinēta*: cf. below n. 13). The jellyfish are referred to again at *HA* 548a10f. as spontaneously generated, and although that does not tell us, of course, that they are not plants, in the context they are associated with other animals, namely various types of testacea.

[11] Peck's additional note, 1970, pp. 352–60, restoring the text with the help of Michael Scot's version, makes the best sense of the passage.

Fuzzy natures?

Pinna, with its byssus.

and muddy places: this *bussos* is the flax-like tuft (hence the name) by which the animal is anchored. (The illustration is taken from Thompson 1947, p. 201). At 548a4ff. it is said not to move and indeed to be 'rooted', *errizōntai*. It is only mentioned in one of our two main texts, but there, at *HA* 588b12ff., it is cited as an example of beings that are difficult to classify as animals or plants on the grounds that they are attached and die if separated, though the pinna is only explicitly said to be attached.

Our final kind, (6), is the razor-shell, *sōlēn*, mentioned along with the pinna in the text just cited, where the *sōlēnes* are said to be unable to live once they are torn off. At *HA* 548a5f. the same point is made, though there the razor-shells (and the cockles) are contrasted with the pinnas in not being 'rooted', *arrizōtoi*. Elsewhere, however, these too are classed as testacea, e.g. *PA* 683b17, and at *HA* 535a14ff. he reports that they are thought (*dokousi*) to be able to hear.

In tackling his problematic cases, Aristotle appeals to a variety of criteria, in some cases, evidently, with rather conflicting results. Four of the main criteria are the following: does the kind have some means of self-preservation? Does it produce residue? Does it have a fleshy part, or more generally otherwise appear to be able to perceive? Can it live 'detached'? It is pretty clear that if the answer to either of the first two is yes, then we are dealing with an animal. In particular he notes explicitly at *PA* 681a33f. that no plant has any residue.

Normally, as we have remarked, the possession of perception is taken to be not just a, but the, decisive criterion to distinguish an animal from a plant, for there are plenty of passages in Aristotle both inside and outside the zoological treatises that suggest as much.[12] But our two texts argue for caution. This is not so much because of *HA* 588b17f. and b26f. The first of those says that some living beings give not a single indication of perception, others only faintly (above, p. 74), but that, we said, might be just a matter of the difficulty in interpreting the indications. The second equally refers to the problem of apprehending (*labein*) any function other than generation, but having said, at b24ff., that the plants appear (*phainetai*: in this case maybe in the sense of what is *evidently* the case) to have no other function, he goes on to refer to some *animals* also (*zōion*, b26). We may compare *GA* 717a21f. which states that with most animals (*zōiōn* again) there is hardly (*schedon*) any function other than, like the plants, producing seed or fruit.

No, the text that gives one pause is rather *PA* 681a17ff., on the third of the kinds we discussed, the jelly-fish and holothouria. They are the subject of a quite unqualified statement: they have no perception. Yet the continuation does not identify them as plants, but says that they live 'as plants that are detached'.

From this it seems that perception is a sufficient, but may not be a necessary condition of animal-hood. Similar doubts may be registered about the fourth main criterion that Aristotle uses, that of 'non-detachability'. If a living being has roots and draws its nourishment from the earth, it is indubitably a plant. But our texts identify a number of problematic cases.

Five of our six kinds manifest, to a greater or lesser degree, non-detachability. Yet non-detachability is no proof that they are plants, any more than detachability is of animal-hood. As to the latter point, he remarks at *PA* 681a20ff. that certain land plants

---

[12] In addition to the passages already cited above (chapter 2 note 69), cf. *de An.* 413b1f. (n.b. *prōtōs*), *PA* 647a21, *GA* 732a12f., *EN* 1170a16. The question of which kinds of animals have which types of perception is the subject of an extended discussion in *HA* IV 8, 532b29–535a27. It is striking that whereas Plato allows anything that has life, plants included, to be deemed a *zōon*, e.g. *Ti.* 77b, Aristotle does not explicitly take issue with him in these texts.

live for quite a time though detached, specifying the rock-plants from Parnassos that do so when hung up on pegs. As to the former point, three of the kinds in question, the sea-squirts, the pinnas, and the razor-shells, while plant-like in their non-detachability, are nevertheless normally treated as testacea and on that score, as also on the grounds of such indications of perception as they give, are animals. Yet that leaves the other two, the sponges and the sea-anemones, particularly problematic since neither of them finds a place either among the testacea or in any other of the main bloodless groups recognised by Aristotle, though by the perception criterion both should be included somewhere. Moreover there is another problem to the application of the non-detachability criterion, namely that this, far more than a specific mode of perception such as touch, is a matter of degree. Those rock-plants could not, presumably, live indefinitely when detached: 'for a long time' (681a24) does not mean for as long a time as they would if they were left on the rocks. Conversely, the sea-anemones that do get detached, *quickly* attach themselves to rocks according to 681b5ff. There are, it would appear, grades of non-detachability, from those things that cannot live at all when detached, to those that are normally quite unattached.[13]

Our dilemma, faced with our two main texts, faithfully reflects Aristotle's. If we follow Aristotle's usual practice elsewhere and treat perception as the sole and decisive criterion for animal-hood, we encounter the difficulty of those creatures that appear to be animals that Aristotle clearly states to have no perception, the

---

[13] This is not the only important differentia that may be a matter of gradation. 'Flying' may be another example. Being able to fly belongs, we are told, *PA* 693b13, to the *ousia* of a bird. However the ostrich ('Libyan sparrow') cannot rise up and fly (*oude petatai meteōrizomenos*) – which leads Aristotle to consider what it shares with quadrupeds, *PA* 697b14ff. Again at *HA* 487b25ff. he uses the term *eupteron* of certain small birds that are good fliers, remarking that conversely they have poor feet (*kakopoda*: cf. the kind called 'footless', *apodes*). As regards the faculty of locomotion (*kinēsis*) itself, Aristotle recognises that some creatures, in their life cycle, change from being motionless, to moving, and back to being motionless again: e.g. *HA* 552a24ff. on *skōlēkes, skōlēkia*, grubs or larvae, and cf. *HA* 551a13ff. on the metamorphoses of the butterfly (cf. chapter 5). Although the nature of the testacea is said to be 'stationary', *monimon*, *PA* 683b5, he goes on to distinguish those that are completely motionless (*akinēta pampan*, cf. *dia telous, de An.* 432b20f.) from those that have 'a little movement' (*mikras kinēseōs*), 683b8ff. We may compare, finally, the degree to which the capacity to continue to live, even when cut up, is manifested by different living things. At *PA* 682b29ff. Aristotle remarks that insects, when cut, may survive for a time, but plants may produce two or more perfect specimens.

jellyfish and holothouria. But to decide to class them as plants, after all, does not exactly fit Aristotle's own language here and it does not tally with the admittedly sparse evidence elsewhere.

One possible response would be to treat these creatures, and maybe others in these texts, as deformities. They must be considered, somehow, as extra numerum. The general classification can stand because these are abnormalities.

It is true that Aristotle is often prepared to label not just individual specimens, but whole kinds of animals 'deformed', judged, that is, by other creatures (often humans) treated as the norm at least with regard to the characteristic he has in mind. The mole is so, compared with sighted animals, because it is blind, e.g. *HA* 533a2ff.; the seal is a deformed quadruped, e.g. *PA* 657a22ff.; the testacea as a whole are with respect to the way they move, *IA* 714b8ff.[14] Yet neither of our two principal texts can be said to make much use of the language of deformity, even though Aristotle does remark that the sea-anemones are incomplete, *ateles*, *PA* 681b6. It is possible that Aristotle would have proceeded in the way suggested, but there is no clear indication that he would. Moreover the consequence would be that the classification would be limited only to what can be considered normal, and that might be thought to be a high price to pay.

The alternative is to countenance the possibility of a variety of criteria to determine what an animal is. That by itself is not at all surprising and it tallies, indeed, with the situation with regard to the kinds of animals themselves, at least insofar as they are to be defined by a plurality of differentiae in the polythetic classification in mind in the criticism of the dichotomists (above, p. 72).

But if no single criterion is both necessary and sufficient, that leaves open the possibility of living beings that meet some, but not all, the criteria in question. The term 'animal' would not thereby become merely equivocal, but the class would have, as it were, a penumbra. Some creatures would meet all the requirements, but others would fail one or more, for example produce no residue: or they might not be fully viable when detached, or they might not have perception.

That too, it might be thought, has a high price to pay, for it suggests a radical break with his usual insistence on the essential

---

[14] I reviewed a number of cases of 'deformed' species in my 1983, pp. 40ff.

status of the perception criterion – and indeed a problematising of the very concept of animal itself. Yet the language of the transition, bit by bit, from plants to animals, combined with the actual analysis of the chief examples he cites, suggests not so much the gradual acquisition of one vital faculty after another, in a clear linear sequence, but rather this penumbral, hazy, effect.

Yet why would Aristotle alter his usual clear view? It is not necessary to think of this being a matter of his changing his mind or of the development of his thought; and indeed that seems very unlikely. Whenever *HA* and *PA* were written, we have found evidence for the more complex view in both texts, not that both treatises do not elsewhere invoke the perception criterion in the way usually treated as canonical.

Rather the situation may be the following: it is only when Aristotle confronts some of the really difficult cases directly that he has grounds for hesitation and (on this suggestion) may be led to modify his usual view. Generally speaking, and for most purposes, the perception criterion will do and can be spoken of as the key determining characteristic. After all the definite exceptions to *that* rule amount just to the jellyfish and holothouria, even though, as he also notes, other animals give no clear indication, or only a faint one, of perception, and in some cases one can grasp no other function besides generation. Yet when faced with those problematic creatures in the sea, Aristotle – on this suggestion – is prepared to revise his idea of what an animal is.

For those who would put a premium on his having an unwavering vision of the distinction between animals and plants, this would be disastrous. But for those who give full weight to the signs of hesitation in a case where he has good reason to hesitate – for as we said it is no easy matter determining the borderline – this is testimony to the flexibility of his work as a zoologist and indeed his open-mindedness.

Imitating Aristotle, I would not be inclined to be dogmatic on the point myself: but I would tend to the latter view.

CHAPTER 4

The master cook

'Concoction' (*pepsis*) is used in an amazing variety of contexts throughout Aristotle's natural science and most especially in his zoology, where it must rank as one of his key concepts.[1] The aim of this chapter is to analyse its diverse uses and in particular to examine them in the light of his programmatic methodological statements in the *Posterior Analytics* and elsewhere. As is well known, considerable importance is attached, in those statements, both to definition and to various modes of demonstration (cf. chapter 1). How does Aristotle's actual use of such a concept as 'concoction' match up to such statements? What, in any event, can be learned about Aristotle's approach to the problems of natural science from an analysis of that use?

Our primary task is the detailed examination of how the concept is deployed in practice. But we should first set out what *pepsis* is said to be, and how its various kinds are specified, in the closest Aristotle comes to a formal definition of them, in *Meteorologica* IV 2 and 3.

Having distinguished two active opposites, hot and cold, and two passive ones, dry and wet, in IV 1, 378b12ff., he begins ch. 2, 379b12ff., by saying that concoction belongs to the hot, and its kinds are *pepansis*, ripening, *hepsēsis*, boiling, and *optēsis*, roasting. Inconcoction belongs to the cold, and its kinds are *ōmotēs*, rawness, *molunsis*, par-boiling, and *stateusis*, scorching, though he immediately notes that these terms are to be taken in a wider sense than their strict usage.

Pepsis is a perfecting, by a thing's own natural heat, from the

---

[1] 'Concoction' is discussed by Peck in 1943 paras. 62–9, pp. lxiii ff. (cf. 1965, pp. lxx ff.), and I briefly reviewed some issues to do with its use in early Greek science in general in 1987, pp. 204ff. However the concept has hardly had the attention in the modern scholarly literature that it might be thought to deserve. Cf. now however Freudenthal 1995.

passive opposites, that are the proper matter for each thing.[2] The principle or the starting-point of the perfecting comes from its own, proper heat, though it can be assisted by extraneous heat as well – as when digestion is assisted by baths. The goal or *telos* is in some cases nature, in the sense of form or substance, in others a realisation of some underlying form, 379b25ff. Examples to illustrate what is concocted and is 'useful' include must, and the pus that gains consistency in tumours or boils.

At 379b32 concoction happens when the matter, that is the wetness, is mastered by a thing's natural heat. Provided this happens in due proportion, this is natural, and such examples as urine, excreta and the residues in general are cited as signs of health when this is so. They are said to have been concocted when their own inherent heat masters the indeterminate. Once a thing is concocted it is thicker and hotter, for the hot makes it more bulky, denser and drier. Conversely inconcoction (*apepsia*) is a lack of perfection (*ateleia*) through a deficiency of natural heat, it too of the passive opposites (380a6ff.).

The following chapter deals with each of the three main species of *pepsis* and their opposites. 'Ripening', in the strict sense, is the concoction of the nourishment in the pericarp (380a11f.) The 'perfecting' here is a matter of the seeds being able to produce another of the same kind. 'Boiling', used strictly of what is boiled in cooking, is 'a concoction by wet heat of the indeterminate material present in the wet' (380b13f.). 'Roasting' is 'a concoction by dry, extrinsic, heat' (381a23f.).

But each of these terms is also applied beyond its strict or primary sense. Thus 'ripening' is spoken of by transfer ('metaphorically', 380a18) in such cases as those of boils and phlegm, when the inherent moisture is concocted by the natural heat. Its opposite, 'rawness', is also 'said in many ways' (380b3ff.), for example of (unconcocted) urine, excreta, 'catarrhs'. 'Proceeding further afield' – as he says, 380b7ff. – (unbaked) clay, milk and other things can be said to be 'raw', if they can change and acquire consistency by heat.

Similarly 'boiling', used strictly of cooking, can be said of anything that contains moisture and that can become denser or heavier

---

[2] πέψις μὲν οὖν ἐστιν τελείωσις ὑπὸ τοῦ φυσικοῦ καὶ οἰκείου θερμοῦ ἐκ τῶν ἀντικειμένων παθητικῶν· ταῦτα δ' ἐστὶν ἡ οἰκεία ἑκάστῳ ὕλη, *Mete.* 379b18ff.

(381a4f.). Here too 'transfer' or 'metaphor' is involved though in the case of the 'boiling' of gold or wood, mentioned at 380b28ff., it seems (if our text is correct) that Aristotle treats them as not the same kind of process.[3] It is however perfectly clear from 381a10f. and b3ff. that both 'boiling' and 'roasting' are natural as well as artificial processes. Thus the concoction of nourishment in the body, that is digestion, is similar to boiling. Both 'boiling' and 'roasting' have their opposite 'inconcoctions', too, for which Aristotle uses the terms *molunsis* (cf. *mōlusis*), 'par-boiling', and *stateusis*, 'scorching'. Here too he is aware of the inadequacies of the ordinary terms available to him in Greek. With 'scorching', for instance, he is clear that there is an inconcoction that corresponds to roasting, when roasting does not take place properly through a deficiency of heat – which happens either because the external fire is inadequate or because of the undue amount of water in what is being roasted. But he introduces this description by remarking that it is 'rather anonymous' (*anōnumōteron*), that is, that it lacks a proper name (381b14ff.).

This brief resumé of *Meteorologica* IV 2–3 already indicates the wide range of applications of *pepsis* and related terms. The focal point of his discussion is that concoction is a bringing of something to perfection or completion achieved through heat operating principally on the wet. Its species differ in the kind of completion they are, that is by the goal or end in view (379b25ff.). They are further differentiated either by the kinds of heat (e.g. 'wet' 'dry') or by the material worked on, and failures, that is 'inconcoctions', are similarly put down to one or other factor or the relationship, the imbalance, between them.

We now face the far more daunting task of analysing how these concepts are deployed in practice in such fields as what we call physiology, genetics, embryology, pathology. Plants provide us with an entry, for the ripening of fruit is mentioned already as one kind of *pepsis* in the *Meteorologica* 380a11ff. *GA* 715b21ff. adds the point that some plants do not produce fruit themselves, but assist in the 'concocting' of that borne by others, instancing the relationship between the caprifig and the fig (a particularly striking

---

[3] The text of *Mete.* 380b30 is disputed. Some editors, with some manuscript authority, omit *ou*, and take Aristotle to be repeating the point made at 380a17f., about 'ripening', namely that the form of the process *is* the same, but the term is used *metaphorai*.

example since this 'assistance', as Aristotle knew, happens through the mediation of insects, the gall-wasp, or *psēn*, described at *HA* 557b25ff.). Then at *GA* 726a6ff. he explains the failure (as he believes) of certain trees, the willow and the poplar, to produce seed, as due to their incapacity to concoct it through their weakness – though their strength is also to blame, since that uses up what would otherwise go to form seed. Finally in *PA* 655b32ff. plants produce no residue and so have no place for it: they receive from the earth their nourishment already 'concocted' – that is compared with the food that animals take in (cf. *PA* 650a20ff.). He means by this that there is no equivalent, in plants, to the digestive processes that go on in the stomachs of animals, though other types of 'concoction' – to produce and ripen the seed and fruit – are of course needed.

With animals the range of modes of concoction grows – as their vital functions do. First there are the processes of digestion, a matter of several stages, with contributions by several different parts of the body, resulting in different end-products. Food, for all creatures, is a matter of the wet and the dry, *PA* 650a3ff., and its concoction and change are brought about by heat. First the food has to be broken up in the mouth, but that is not responsible for *pepsis*, but rather for *eupepsia*, 650a10ff. That is, the success of the process depends on mastication – for when the food is reduced to small pieces it is easier to digest – though mastication is not itself concoction.

The first phase of concoction proper happens in the stomach, which produces on the one hand 'useful' residues, and on the other 'useless' ones, a distinction that corresponds, according to *GA* 725a3ff., to that between what is useful and useless in the nourishment. The useless residues get voided, though if the useless elements in the food are in excess, or are insufficiently well concocted, they can cause diseases. But if they are in due proportion and duly concocted, they are naturally excreted and, as the *Meteorologica* had already suggested, 380a1, are signs of health.

However it is not just the stomach that is involved in concocting the food. It passes on what it has digested to those blood-vessels, *phlebes*, that start at the base of the mesentery, according to *PA* 650a28ff. The most important useful product of digestion is the blood, said to be (*PA* 650a34f., *GA* 726b1ff.) the final form of nourishment for all the blooded animals – while in the bloodless ones what is 'analogous' to blood serves this role (cf. chapter 7). Blood is produced primarily in the heart, *PA* 647b4ff., *HA* 521a9f., which is the source of all the blood in the body, *PA* 665b14ff., 666a6ff. Again, things can go wrong. If the blood is insufficiently concocted, the result is *ichōr*, serum, *HA* 521b2f., *PA* 651a17ff., or else watery blood, *PA* 668b8ff., even though one passage, *HA* 521a17ff., suggests that blood may itself be concocted from *ichōr*.

The production of blood is, one may say, the key step in the whole process of nutrition, for blood does not just nourish the other parts of the body, *PA* 650b8ff., 652a6ff.; it is the material from which all the parts of the body are formed, first the heart itself and then all the other viscera, *PA* 665b5f. Every one of the parts somehow comes to be from the blood as it is concocted and divided up, as he puts it at *GA* 726b5f.

Moreover in addition to the heart, other parts of the body too help in various ways with the concoction of the nourishment. The liver and the spleen have this role (*PA* 670a20f.), indeed the liver is said to be for the sake of concoction (a27). Similarly the kidneys help in the concoction of the useless residues, *PA* 670a22ff., 672a20ff., thanks to the heat they contain. The kidneys themselves are fat, since the blood that is left in them – after the process of the filtering of the residue – is *eupepton*, easily concocted, and when there is *eupepsia* of the blood, the end-result is lard or suet (*PA* 672a1ff.). This leads Aristotle to draw an analogy between burnt solid substances and concocted liquids. In the former, in ash, for instance, a certain amount of the 'fire' is 'left behind'. So in concocted liquids a part of the heat that has been generated is left behind. He uses this to account for oily substances coming to the surface of fluids (presumably because of the heat they contain) and he then applies that to the formation of fat on the outside of the kidneys. In some animals it takes the form of lard (*pimelē*: soft fat), in others suet (*stear*: hard fat). He had explained these as differing according to the quality of the blood from which they are concocted (*PA* 651a20ff.). Both consist of blood that has been concocted as a result of an abundance of nourishment, when the surplus, being well-concocted and well-nourished, is not used up simply to nourish the fleshy part of the animal.

Marrow, similarly, is suety in those animals that produce suet and lard-like in the others (*PA* 651b28ff., 652a7ff.). In both cases, marrow too comes from concocted blood, but with marrow there is a further factor, the heat produced by its being surrounded by bone. Although brain is not made of marrow (as many are said to think) and is indeed opposite to it in its nature, being not hot, but the coldest of the parts of the body (*PA* 652a24–9), it too is the result of a process of concoction. The concoction it requires, however, given its nature, is correspondingly more extensive (*GA* 744b1ff.). At *GA* 744a16ff. he speaks of a process of evaporation and concoction that contributes also to the development of the eyes and their change of size as the embryo grows.

Both the formation, and the function, of the omentum, *epiploon*, discussed in *PA* IV 3, 677b14ff., involve concoction and provide a clear illustration of the dual deployment of that concept. The omentum is a membrane formed of suet or lard as the case may be. It is the necessary product of a process that Aristotle describes. When a dry and wet mixture is warmed up, it always forms a skin-like membrane. The surrounding region (near the stomach) is full of nourishment. Because of the density of the membrane the bloody nourishment that percolates through it must necessarily be fatty. But because of the heat in that part of the body it will become concocted together (*sumpettomenon*) and form suet or lard instead of some fleshy or blood-like structure. So the omentum is itself the result of a distinctive process of concoction, like the other viscera. However, nature uses it to advantage for the *eupepsia*, ease of concoction, of the food, that is to help the animal concoct its food more easily and more quickly. It does this because it is fat, and the fat is hot, and it is (of course) the hot that brings about concoction. Itself the result of concoction, it contributes in turn to the concoction of other things.

We have noted that things can go wrong. The useless residues can give rise to diseases, and *ichōr* can be the result of the inconcoction of the blood. Thus *pepsis/apepsia* figure prominently in Aristotle's accounts of many pathological complaints and of the debility or weakness of the young and old as well as of the sick.

Thus at *PA* 668b1ff. he first explains sweat as the percolation of the residue of moisture through the smaller blood-vessels when the body is thoroughly heated. In some cases, however, the sweat itself is a blood-like residue, when the body is in poor condition, due in part to the blood being insufficiently concocted and watery – that is, it has not been thickened properly, for he thinks of thickening as one effect of *pepsis*. Here, as elsewhere, the cause of the problem may be either insufficient heat, or too much material for it to work on (668b11ff.). Other kinds of morbid bloody discharges are also attributed to inconcoction in the blood-vessels, just as 'diarrhoea' occurs through inconcoction in the intestines (*GA* 728a21ff.).

Similarly a variety of signs of weakness, from grey hair, to eyes that are *glaukos* (blue-grey) in colour, is explained in terms of inconcoction, in both those instances of the moisture in the brain, *GA* 780b6ff., 784a34ff., b11ff. In the second of these texts he refers to a now lost work, *On Growth and Nutrition*, for a detailed account to support the generalisation that in every part of the body, damage and disease result from the failure of its proper heat to concoct the nourishment that reaches it. More generally, old age, thought of as cold, is said to be a kind of 'natural disease', while disease is an 'adventitious old age', *GA* 784b32ff., while with that cooling goes, of course, an increasing inability to concoct, e.g. *GA* 725b21f.

The next main area of application of the concept of concoction that we should review is in the accounts Aristotle offers of the differences between different kinds of animals. Concoction is a natural process, or rather a whole variety of them. It results from a creature's own proper heat and so one might have thought that every species of animal would perform the concoctions it needed in the appropriate fashion. But that is to discount that, in Aristotle's resolutely hierarchical scale of beings, some animals do things very much better than others. Some may manage satisfactorily only by dint of some special adaptation to counterbalance a potential weakness (not an adaptation associated in any way with any process of evolution). Sometimes that counterbalancing still leaves the creature with certain deficiencies.

Take what is said of the organs of digestion in a variety of quadrupeds, birds and fish in *PA* III 14. Some animals, such as the camel, have several stomachs as they have to deal with food that is not easy to concoct, being thorny and woody, and so too do the horned animals that have front teeth only in one jaw (that is, the lower) (674a28–31). Later he points out that the pig's stomach, for example, has folds for a similar purpose, to prolong the time of concoction (675a27ff.).

The birds, by contrast, have no teeth, so the service they perform in the preparation of the food has to be done differently. In some cases, this happens in the 'crop'. In others, the stomach is strong and fleshy so that it can retain and concoct the food in its unmasticated state for a considerable time (*PA* 674b26ff.). Here nature compensates for the deficiencies of the mouth by the power and heat of the stomach.

Fish too have similar problems. They have teeth, but do not spend long chewing (674b34ff.). To assist in the process of concoction, some have fleshy stomachs, like those of birds: most have appendages by the side of the stomach in which they can store the food (as if in storage pits) and macerate and concoct it, though these appendages are different from those of birds, being high up near the stomach, not low down at the end of the gut. Because of the deficiencies in their digestive organs, fish as a whole are gluttonous. What happens is that the food passes through them (comparatively) unconcocted, especially in the case of those with a straight gut. So although the fishes' digestive processes are helped by certain special anatomical features, shortcomings evidently remain, for which the creatures have to compensate by eating more often.

Nor is it just the differentiation of the primary organs of digestion in different animals that is associated with the relative ease or success in concoction. The sea-urchin's spines are another case, discussed at *GA* 783a19ff. Those that live in deep water are small but have big, hard spines. They are big because the growth of the body is diverted to them. As the creatures have little heat and are unable to concoct the food, they have a good deal of residue (that is unconcocted residue) and it is from this residue that the spines are formed. *PA* 681a2ff. further elaborates on the differences between types of sea-urchin, saying that those with larger eggs have a hotter constitution. The inedible varieties, with smaller eggs, are full of residue, where this is not just a matter of their external spines. That is in line with what is said elsewhere, where, for instance, the meat of white animals is said to taste better: they are hotter and so more efficient at concoction, and that makes them sweeter, since concoction produces sweetness (*GA* 786a15ff.). Similarly unfertilised 'windeggs' are less good to eat and less sweet, since less concocted, than ordinary, fertile eggs (*GA* 750b24ff.). Again it is only when the foetus is fully developed, and the concocted food is no longer diverted to it, that the milk the mother produces becomes sweet (*GA* 776a25ff.).

I have left until last the complex of ideas relating to generation and reproduction that constitutes what is probably the most important of all the fields of application of the concept of *pepsis*. There are no fewer than five principal contexts (as well as a fair number of minor ones) in which concoction is involved in Aristotle's account of the reproduction of vivipara: (1) The production of semen; (2) that of menses; (3) the action of the former on the latter; (4) the mother's contribution to the embryo's concoction; and (5) the differentiation of the parts of the growing embryo including its success or failure in taking after its parents. The corresponding processes in ovipara make equally heavy use of the same family of concepts, and so too do other aspects of Aristotle's theories of generation all the way to his account of spontaneous generation itself.

(1) We have already seen how the concoction of nourishment produces not just blood but a number of other homoeomerous substances in the body as well, such as lard, suet, marrow. Semen itself (*gonē* or *sperma*) is also a residue of the useful nourishment, indeed in its final form, *GA* 725a3ff., 726a26ff., b5ff. The role of concoction in its production is mentioned at 725b21f. (the old produce no semen as they are unable to concoct it) and at 726a3ff. where it is compared with fat and indeed connected with it: fatter people produce less semen. Thus far an analogy with plants' seed (also *sperma*) holds well enough, and indeed plant comparisons punctuate his account of generation, e.g. *GA* 715b16ff., 731a1ff.

But in most animals, unlike plants according to Aristotle, male and female are distinguished, the male being responsible, as he puts it, for the form and efficient causes, the female for the material, e.g. *GA* 716a4ff. But male and female are defined by what they can or cannot do, by a capacity or an incapacity, and in particular by what they can do with their useful nourishment, *GA* 765b8ff., 766b12ff. The males, being hotter, can concoct it, in its final form, into semen, (e.g. 719a35ff.). Females, being colder, fail to do so and produce instead the menses, a kind of impure, unconcocted, seed, *GA* 728a26f., 737a27ff., 774a2f.[4]

(2) So females are defined, notoriously, in terms of this deficiency, as a kind of natural deformity, 737a27ff., 775a14ff. Despite

---

[4] At *GA* 728a17ff. females are compared with young, infertile, males from the point of view of the failure to concoct semen.

the fact that Aristotle is well aware that several of his predecessors, including for example Parmenides, had argued – from the evidence of the menses, indeed – that females are hotter than males (e.g. *PA* 648a28ff.), he maintains they are colder. Indeed at *GA* 765b8ff. he uses the theory that they do not concoct seed, viz. in the way males do, as *evidence* that they are colder, because concoction is achieved by heat. However, at *GA* 726b30ff. we find him arguing the converse, that is *from* the nature of women as weaker and colder *to* the conclusion that they produce more residue, but less concocted residue, that is 'a quantity of blood-like fluid'.

(3) What happens when the semen encounters the menses is described metaphysically in terms of an interaction between the form and efficient cause on the one hand, and the matter on the other. Yet that masks the detail of the interaction. At *GA* 729a16ff. (cf. 730a15ff.) the work of the semen is to concoct the material, in a way he compares several times to the action of rennet curdling milk. But that is a closer analogy, for Aristotle, than one might at first sight suppose. The matter, in both cases, the milk and the menses, is a residue of useful nourishment, and rennet, described as 'milk that has vital heat' (739b22f.) shares with semen that it can 'set' the material it works on thanks to this quality. The analogy also assists Aristotle to the conclusion that the semen does not become *part* of the developing embryo, for he thinks of the agents that bring about curdling as acting *just* as efficient causes, *GA* 737a12ff. But the 'setting' of the menses by the semen is a 'concocting' of it.

(4) Moreover it is not just the case that the female provides the matter, as well as the place, for the new embryo. Although females are unable to concoct semen for themselves, they have two important other contributions to make. First, as noted, they concoct the milk that is the nourishment of the new-born offspring. Then also the womb is relevant to the embryo's concoction. At *GA* 719a33ff. Aristotle discusses the reason for the womb being inside and says that this is due to the embryo needing shelter and concoction – while the outside of the body is easily damaged and cold. The womb is not itself said to concoct the embryo, and the contrast between its role and the stomach's concocting food is forcefully pointed out on several occasions, *HA* 538a8ff., *GA* 756b10ff., b27ff. Yet the womb's position is explained in part in terms of the needs the developing embryo has for *its* concoction.

(5) Finally there is that concoction itself, namely that involved in the growth of the embryo and the differentiation of its parts. That differentiation, according to *GA* 775a16ff., is a separating out, *diakrisis*, which may happen faster (in the case of male embryos) or slower (females). But this separating out is a concoction, *pepsis*, brought about (as always) by heat. This enables Aristotle in that passage to explain why in humans, male embryos are subject to more accidental deformities in the womb: they move about more (*GA* 775a4ff.). Yet once females are born, they develop more quickly, reaching maturity and old age faster than males – which Aristotle puts down to a quite different principle, namely that inferior things can be brought to an end or goal (*telos*) more quickly (775a20f.). Differentiation in the womb, it seems, does not follow the same pattern as the maturing of the animal once born. But that he thinks of the former in terms of concoction can be seen also in such other parts of his account as that of the formation of the eyes and brain (744a16ff., b1ff., above p. 88) and in his more general reference to the young of viviparous selachians (cartilaginous fish: dogfish, sharks and rays) being concocted and perfected in the womb, *HA* 565b22f.

While the vivipara provide the models of what animals should be, Aristotle deals at length also with a variety of problems that arise in oviparous generation. How do wind-eggs differ from fertilised ones? They are far more unconcocted, *GA* 750b24ff. Eggs show that the female can, up a point, produce by herself: but she needs the male principle to *re*produce, *GA* 730a29ff.

Then what is the difference between the two parts of the egg, the yolk and the white, *GA* 753a34ff? They have opposite natures, as one can tell from their reactions to frost or heat or fire, but both are subject to concoction. The yolk is so when the egg is incubated, and it then becomes the nourishment for the embryo chick. But the white is also 'concocted' – to become the chick itself (753b2, b9f.).

The process of hatching, accordingly, is one of concoction, in part by the bird (especially the female, 752b15ff., 753a7, but cf. *HA* 562b17ff.), in part also by the earth or by warm weather, 752b29ff. Indeed with some oviparous quadrupeds it is the heat of the earth that contributes principally to the concoction: the parent animals visit the eggs and incubate them rather for the sake of protecting them, 752b32ff. There is a difference here between oviparous quadrupeds and birds: the eggs of the latter need the hen's incubation, while those of the former are strong enough to be fully concocted by the warmth of the climate alone, 753a5ff. In either event, hatching is faster on sunny days.[5]

However balance, as usual, is needed. Excessive heat spoils the eggs, 753a21ff., just as indeed in viviparous reproduction there has to be due balance, a *summetria*, between the male and female factors. That is necessary first for any successful reproduction, 729a16ff.: but Aristotle is led to offer an elaborate account of children taking after one parent or the other, not just as male or female, but in physical characteristics, in *GA* IV 2–3, where the balance between male and female principles is a chief factor in achieving the ideal. Thus male children taking after their fathers reflect in part the well-concocted quality of the female spermatic residue, *GA* 767b15ff., and departures from that ideal, the failure of the male principle adequately to 'master' the female, can be due either to a deficiency in its power to concoct, or to the abundance of the material there to be concocted, 768b25ff.

As one declines further down the scale of being, the extraneous heat, already referred to in the reproduction of ovipara, takes over more and more of the role of the heat otherwise provided, in one form or another, by the parents. Thus after copulation the female grasshopper deposits the egg-like grubs or larvae directly on the ground, or rather just below its surface, in a honeycomb-like structure compared to a membrane. There they undergo 'full concoction', but for this to happen is has to be summer, for in winter the eggs stay dormant (*HA* 555b18–556a7).

In spontaneous generation itself (to which we shall be returning in chapter 5) there are no parents as such at all. Everything depends upon the way certain putrescent matter is acted upon, whether putrescent earth or residues, *GA* 715a24f. (cf. b26ff., the same thing may happen in plants).[6] However, as *GA* 762a13ff. explains, nothing is produced by putrefaction in itself, but rather by a process of concoction, *pettomenon*. The putrefied matter is it-

---

[5] At *GA* 766b34ff., *HA* 573b34ff., he endorses the popular belief that male offspring are more likely to be conceived when the winds are northerly: when they are southerly, more residue is produced and it is harder to concoct.

[6] *HA* 550b32ff. specifies a wide variety of materials from which animals may be spontaneously generated, including dew, putrefying mud and dung, wood, hair, flesh, residues (whether or not already excreted); 552b6ff., 10ff., add snow and fire, 557b1ff. wool, clothes and books: see further ch. 5.

self a residue of what has been concocted, that is, by an earlier process of concoction. But as in regular reproduction, so too in spontaneous generation, what is needed is not just the right material (this previously concocted putrescent matter) but also heat to bring about its concoction. This is provided principally by warm weather, *GA* 743a35f., 762b12ff., 14ff.,[7] the heat of the sun performing, in these cases, the role that the heat of the parent animals does in others, *GA* 737a3ff. Thus although the lack of a species-specific efficient and formal cause presents a major *difference* between spontaneous, and other modes of generation, what *links* them all, in Aristotle's account, is the idea that new life comes to be as the result of processes of concoction.

The great strength of Aristotle's use of the idea of concoction lies, in general, in the way it enables him to see the connections between widely disparate phenomena and processes. But the corresponding weakness is in the very vagueness or generality of the concept – which is what allows him to suggest those connections. To put it another way, the connections he apprehends run ahead of the theoretical explanations he can offer.

First there are problems to do with the relationship between the detailed applications of the concept and the general account he sketches out in *Meteorologica* IV and more especially with the differentiation of different types of concoction and their end-results. Then we must take up, in conclusion, the question of what this example can teach us about Aristotle's practice of scientific inquiry as opposed to his programmatic methodological statements on that subject.

Let us turn back first to the explicit account we are offered of what *pepsis* is at *Meteorologica* IV 2 and 3. It and its species are a perfecting, by heat, of certain materials, primarily of the wet. Even when we add some specificities as to the types of heat ('proper' 'natural' or 'extrinsic', 'dry' or 'wet') and note that the matter is the proper matter of the thing in question, the inadequacies of this as a *definition* are obvious, whether by Aristotle's standards of definition or our own. What kind of 'perfecting' is in question? This will turn out to be a key issue. But presumably not

---

[7] *HA* 551a2ff. recognises that spontaneous generation can also occur in winter, after a spell of fine weather.

just anything that reaches a *telos* will count – for otherwise that risks collapsing the whole account into one of the final cause, reducing it to the observation that final causes are at work in physical changes.

Again when the different types of heat are mentioned, in what lies their difference, *qua* heat? How is the wet, or the wet and the dry, to be understood? Are the differences in the end-results – from the concoction of semen to that of boils – to be put down to the nature of the heat, that of the wet/dry, the place where the concoction occurs, or the mode of 'perfecting', or to all of these? Would any kind of interaction of hot and wet that ends in something useful do? Surely not: at least there is no evidence to think that Aristotle would accept the fusion of two metals (that is, what he would call a *mixis*) under the action of heat to produce a new alloy as an example of *pepsis*. Yet in that example all three criteria would appear to be met, the role of heat, acting on the 'wet' (metals being classified as such), to produce something useful.

Many of the problems stem from the vagueness of 'hot' 'cold' 'wet' and 'dry' themselves. Aristotle recognises that these are 'said in many ways', especially at *PA* II 2 and 3, for example at *PA* 648a36ff., 649b10ff. But there, while he acknowledges just how disputed these terms were among his predecessors and contemporaries, and points to important differences between, for example, innate, and acquired, heat and between what is hot in potentiality, what hot in actuality, he does not resolve the major difficulties, not offer any clear conclusions, there, as to what precisely hot, cold, wet and dry are. They are repeatedly invoked in the explanations of other things: yet *their* nature to some extent remains unclear.

That remains the case even when he does give what some have taken to be a definitional account, when he considers their role as the primary qualities of his element theory, in *De Generatione et Corruptione*, 329b26ff.[8] He there takes 'hot' to be 'that which combines things of the same kind' (τὸ συγκρῖνον τὰ ὁμογενῆ). 'Cold' is 'that which brings together and combines homogeneous and heterogeneous things alike' (τὸ συνάγον καὶ συγκρῖνον ὁμοίως τά τε

---

[8] Cf *Mete.* 378b20ff., which speaks of defining (*horizometha*) the natures of hot, cold, wet and dry, the first pair as active (as *sunkritikon*) the second as passive (in terms of *euoriston* and *dusoriston*).

συγγενῆ καὶ τὰ μὴ ὁμόφυλα). 'Wet' is 'that which, being readily delimited [i.e. by something else], is not determined by its own boundary' (τὸ ἀόριστον οἰκείῳ ὅρῳ εὐόριστον ὄν), and 'dry' is 'that which, not being readily delimited [i.e. by something else], is determined by its own boundary' (τὸ εὐόριστον μὲν οἰκείῳ ὅρῳ, δυσόριστον δέ).

The ambition, here, to give a general characterisation of the four fundamental qualities is clear. Yet they are hardly adequate as definitions. This is not just that they are not given canonical *per genus et differentiam* form: nor could they be reached by standard Aristotelian techniques of division. Rather the chief problem relates to the mismatch between definientia and definienda. While the contrariety of 'wet' and 'dry' is reflected in their characterisations, that between 'hot' and 'cold' is less clearly so in theirs. Indeed both 'hot' and 'cold' *share* the capacity to combine things (*sunkrinon*), even though there are differences in the specification of what they combine. Moreover a capacity to combine things, whether of the same or of different kinds, is shared, one might think, by many other types of agents, and in the *Physics sunkriseis* and *diakriseis* are treated as local movements, *kinēseis kata topon*, 265b19ff., cf. 243b7ff.

Moreover so far as the usefulness of these characterisations goes, in relation to the definitions of *pepsis* and its species, the fundamental difficulty remains that, given that *any* physical object is either hot or cold, either wet or dry, the specification of *pepsis* as an interaction of hot and wet remains thus far indeterminate, if not totally vacuous.

The very variety in the objects and processes to which we find *pepsis* applied carries with it its own problems, of two main types, relating (1) to the different operations of *pepsis* within the same kind of animal, and (2) to its apparently similar operations in different species.

Take first the first type of difficulty. How are we to account for all the different end-products described as the results of 'concoction'? Why, in some cases, does it give rise to one kind of homoeomerous substance, say fat, in others another, say lard or marrow or again milk or semen – all of these described as due to the concoction of the blood? Why do the processes of concoction at work in the body lead now to the formation of the omentum, now to the brain? In some cases the texts of Aristotle yield an answer, and in others one can be inferred or guessed: yet if we press this type of question the difficulties he appears to be faced with are considerable.

In some cases we may invoke the place where concoction takes place as a relevant factor, the kidneys, say, or the heart (cf. above p. 87), and elsewhere differences in the end-result can be put down to differences in the blood, so provided these can be explained (see below), that will serve as an account of these differences too. Again one of the distinctive features of the production of marrow is the extra heat supplied by the surrounding bone (*PA* 652a7ff., pp. 87–8).

But the problem is not just a matter of why different parts of the same animal come to be produced – mainly from its blood – but relates also to how the same part in different animals does, the second type of difficulty I mentioned. All the vivipara produce semen, and no doubt we can say that the distinctive characteristics of a tom-cat's semen or a dog's (that enable them to act as the efficient and formal cause in the generation of new young cats and dogs) reflect the differences in those animals' blood. Blood is blood, of course, in any creature: but Aristotle engages in a careful discussion of the differences between different animals' blood in *PA* II 2–4. As we have noted before (chapter 2 at pp. 44–5), these form part of his account not just of the 'bodily' but also of the 'psychic' characteristics of the animals concerned.

Thus far the account seems readily intelligible. But the question we have to pose is what room Aristotle's theory of concoction allows for those differences in the blood. Blood, after all, is the product of the concoction of the useful nourishment. But the nourishment of different kinds of animals may (in certain cases) be identical – and yet the blood they produce, and what it produces in turn, differ markedly. Consider the case of various creatures for which cow's milk may be food. When the cow's own calf consumes the milk, it concocts the blood that is characteristic of cows and it grows. But when humans or cats drink the same milk, that milk eventually contributes to the very different result of the humans' or the cats' growth, mediated, of course, by the process of turning the cow's milk into human or cat's blood as the case may be.

Obviously something more than the matter worked on must be involved. Yet it is not as if Aristotle tells us that it is because of the differences between our stomachs and the cat's that the blood turns out different, even though, as we saw (pp. 89f.), differences between the stomachs of different species do occupy his attention. But when they do so, the focus of his interest is rather on how efficient their concoction is, and what particular anatomical features are necessary to secure that efficiency, not on how the distinctive end-results – the different kinds of blood – are produced. He is concerned with how good a digestive organ a stomach of a particular kind is, not with how it manages to turn, say, *cow's* milk into *human* blood.

But Aristotle's seeming indifference to this second type of objection reflects a more general feature of the styles of explanation he attempts. He works back, one might say, from the end-result to its necessary conditions, its material and efficient causes.[9] Of course he is also intensely interested in characterising the forms and in identifying the final causes at work. Yet from certain points of view, at least, the forms, that is, the specificities of the end-results, are taken as givens.

This is analogous to the 'top-down' style of account that he favours in areas of his psychology.[10] Certainly he does not there attempt to build up explanations of the vital functions from their material constituents, as if it were a question merely of those constituents acquiring the new functions as they become more compounded. Those vital functions belong to the essences of the animals in question – to their natures – and the natures determine the matter, not the matter the natures.

So too in the account of concoction, the key to one of the difficulties I pressed (to do with milk being turned into different blood in a calf, a cat and a human) would appear to be the – unexpressed – point that the 'perfecting' that concoction *is* is species-specific. It is because the animal is the kind of animal it is that the modes of concoction it exhibits produce the *particular* results they do.

We have two points to balance then. On the one hand digestion is digestion, wherever it occurs, and so too with semen-production and the whole gamut of processes we reviewed: that is where the concept of concoction enables Aristotle to *generalise* across species,

---

[9] One may compare what is said of the mode of demonstration proper to the study of what comes to be naturally at *PA* 639b30ff., on which see above ch. 1 at pp. 29ff.

[10] Cf. above ch. 2 at p. 64.

including across widely divergent species, as when he can consider the spontaneous generation of many different kinds of creature as a kind of concoction *like* other modes of generation (cf. chapter 5).

On the other hand, and in this context more importantly, the effects of digestion in one animal will differ from those in another (the blood will vary) and *these* differences will depend on the form or the essence of the animal in question. So too, even more vividly, in the case of the concoction of semen, where the results of the same *general* process yield young of each kind of viviparous animal, and so too with the ovipara and the rest. Yet if that invoking of essences helps to meet one type of difficulty I raised (to do with species-differentiation), we shall have to return shortly to the other, to do with the differences in the modes of concoction themselves, that is between concoction as digestion, concoction as organ-formation, concoction as semen-production, as foetus-differentiation and the rest.

Neither the general account of *pepsis* in the *Meteorologica*, nor the actual highly diversified use of the concept in zoology and elsewhere, fits at all easily into the ideal programme for science developed in the opening chapters of the *Posterior Analytics*. This insists, as far as possible, as we have seen (chapter 1), on rigorous axiomatic-deductive demonstration, where definitions figure among the ultimate indemonstrable but self-evident primary premises. The validity of syllogistic reasoning as a whole depends on the terms used being univocal, at least over a given stretch of argument – and if the reasoning is to be part of a fully worked out science, then over *all* the relevant stretches of argument in the field. The possibility of *APo.* style demonstration further depends on the additional requirement of the terms having explicit definitions.

By those criteria, the term *pepsis* must be thought to rate very poorly indeed. The account given in the *Meteorologica* is highly indeterminate, and it allows, even positively exploits, transferred as well as strict uses. When Aristotle finds himself 'proceeding further afield' (*Mete.* 380b7ff., above p. 84), this comes close to acknowledging what I call 'semantic stretch'. Moreover, in practice, in his actual use, the term's polyvalence is very marked. Concoction as digestion, as organ- or residue-formation, as semen-production, as foetus-differentiation, as the hatching of eggs, or in the complex manifestations of spontaneous generation, all *count* as 'concoction'.

Aristotle sees *something* in common to all these phenomena. Here the variety cannot be put down (just) to the differences between the essences of different species, between the form of a cat, as it might be, and the form of a human. Rather, we have first to complexify the notion of what the essence of each species may cover, the variety of its vital activities. Then beyond that, we have to recognise the connection not just between apparently very diverse vital processes within the same animal, and again across the animal kingdom, but between all of those and the different processes exemplified by plants – in their seed-production and the ripening of fruit – and further still to the modes of concoction that take place outside living organisms, such as the boiling or roasting of food. Yet his tentative apprehension of links between all of these phenomena only seems possible thanks to the indeterminacy of what will count as some kind of 'perfecting', by some kind of 'heat', of some kind of 'wet', in some kind of organ or suitable environment. The very fertility of his notion of concoction is, one might say, a function of its indeterminacy.

But would Aristotle have done better to have conformed more closely, in his use of *pepsis*, to the models and ideals that the *Posterior Analytics* provides? Does not the *APo.* itself envisage zoological and botanical examples alongside the strictest mathematical ones? Should he not have worked harder at reducing the polyvalence of 'concoction'?

First, as to the application of *APo.* notions of definition and univocity to animal and plant examples, it is clear from the illustrations given in book II especially that the model of demonstration elaborated there is not confined to mathematics or to such astronomical examples as the explanation of eclipses (see below chapter 8). Nevertheless I have already expressed reservations (chapter I at p. 13) about the actual zoological and botanical examples given, which seem in some cases illustrations not of actual demonstrations, but rather of the type of reasoning suggested. This certainly applies to the well-known botanical example begun in II 16 and continued in the next chapter. At ch. 16, 98a38ff., we are first offered 'broad-leaved' as the explanation of 'deciduousness' (though these terms convert); then at 98b36ff. the 'coagulation' (*pēxis*) of the wet (*hugron*); then in the following chapter, 99a27ff., the 'coagulation' of the 'sap' (*opos*) 'or something like that' (*ē ti allo toiouton*). The ambition to provide strict demonstrations in botany too is clear. But quite apart from the problem of whether such terms as 'coagulation' and 'sap' themselves can be given rigorous definitions (when they are used in the definition of 'deciduousness'), the addition of the phrase 'or something like that' indicates that this is hardly presented as a definitive actual demonstration.[11]

But should not Aristotle have worked harder to implement the recommendations of the *Organon* in his actual zoological practice? Is the kind of indeterminacy present in his use of the concept of *pepsis* tolerable as science?

To answer those questions we have to consider what the effects would have been if *Posterior Analytics* standards of rigour had been demanded of *pepsis*. What they required was strict definitions that could serve as the middle terms of demonstrations, linked, ideally, in a comprehensive axiomatic-deductive network. Yet that would certainly have inhibited, even ruled out, the type of wide-ranging application of the concept of *pepsis* that is characteristic of Aristotle's use. Indeed what permits that range of application is, precisely, the indeterminacy that conflicts with the *APo.* requirements of univocity. The indeterminacy we find in his use is precisely what allows him to suggest possible connections, as we said, not just between digestion in one kind of animal and in another, but between digestion and cooking and organ-formation and semen-production and fertilisation – and all the rest.

From the stand-point of some theories of science – from that of the *APo.* itself indeed – indeterminacy in any form is intolerable. But scientific inquiry is, of course, not just a question of being able to give strict demonstrations, employing rigorous definitions. This is not just a matter of adding in the topics of discovery, or of hypothesis-formation, to that of demonstration, but also of allowing an open-endedness in the exploration of issues at a stage when it may not be entirely clear precisely what the issues – the scope of the explananda – are. In that context the types of concerns that preoccupy the *APo.* are irrelevant.

But if that kind of open-endedness is detectable in his actual use of the concept of *pepsis*, where does that leave his practice in relation to the ideals set out in the *Posterior Analytics*? Of course

---

[11] Indeed the discussion of why plants lose their leaves which Aristotle gives at *GA* 783b10ff. proceeds rather differently, in terms of a deficiency (not a coagulation) of fatty fluid (not sap) – not that it is offered as a syllogistic demonstration in the manner of the *Posterior Analytics*.

the fact that the practice may, in certain respects, conflict with the ideal does not show that he has modified that ideal. Even if, where 'concoction' is concerned, the ideal, I would argue, is not just not applied, but not applicable, that does not mean that Aristotle *thought* it was in principle inapplicable. On that topic, too, indeterminacy is inevitable. Yet his actual practice suggests that he was less concerned with striving for univocity than with exploiting the stretch of the concept, with exploring the similarities between the very different types of phenomena and processes he encompasses under the same polyvalent rubric. For that exploration, the polyvalence of the rubric, so far from being a shortcoming to be lamented, a blemish to be removed, is a positive advantage: and my inclination would be to say that Aristotle saw it that way too.

Pepsis, concoction, was called the 'maister Cooke' in Spenser's Faerie Queen, a tag that Peck used as the epigraph for his edition of the *de Partibus Animalium*. But in his exploitation of the polyvalence of the concept, it is, one might say, Aristotle who is the master cook.

CHAPTER 5

Spontaneous generation and metamorphosis

Aristotle said that wonder is the origin of philosophising, *Metaphysics* 982b12ff., but he was not too specific, in that text, about what we should wonder at, nor did he make much allowance for the possibility that although, as he put it, wonder can lead people to desire to know, it often does not. We are often content to amaze or be amazed without inquiring into the causes or explanations of the strange objects or phenomena in question, a tendency that can be illustrated readily enough in many ancient collections of *Mirabilia*.

What was there to wonder at in the animal kingdom? It is not as if the Greeks all had the same ideas about that. On the contrary, it was a matter of some dispute, since the reaction of some to what others believed amazing was flat disbelief. They were in business to redefine what was truly amazing, whether or not they then presented themselves as being in a position to offer explanations of that. If they did, then they scored twice over, both knowing what was truly wonderful, and knowing how to reduce it, too, to the intelligible.

Many aspects of the generation of animals and of the changes they undergo occasioned no surprise. But animals that were said to arise 'spontaneously', or that underwent radical changes in their form during their life cycle, could. By 'spontaneous generation' can be meant different things.[1] In Aristotle's *Physics*, 196b10ff., 197b18ff., 198b34ff., the spontaneous (*to automaton*) picks out what does not happen 'always or for the most part', and chance (*tuchē*) refers to one species of the spontaneous, namely those effects that could have been produced intentionally, and are the results of action by intentional agents, but were not produced as the result of their

---

[1] There is a clear account of the differences in Balme 1962b.

particular intentions on the occasion in question. On this view the spontaneous is the unusual, the exceptional, as opposed to the natural as defined, precisely, as what happens 'always or for the most part'.

This does not correspond to the principal use in Aristotle's zoology.[2] There the spontaneous may refer simply to what is produced without human intervention, as when health comes to be without the aid of a doctor (*PA* 640a28f.). But in the context I am chiefly interested in, in certain discussions of spontaneous generation, that is defined as those cases of generation where the creatures produced either have no parents as such at all, or at least are not of the same kind as the offspring (*HA* v 1, e.g. 539a22ff.). Many of the animals Aristotle discusses are said to come from mud, or earth, or putrefying matter: we shall be commenting on this later. But at *GA* 732b11ff., *HA* 539b7ff., for instance, he recognises some cases where the animals in question are male and female: they copulate and produce something, but what they produce is incomplete and the animals themselves come to be 'spontaneously'. Again he also counts the production of eggs without copulation by birds or fish as the 'spontaneous' production of those eggs, *HA* 539b2–7.

The most important point that will emerge from our analysis is that the creatures 'spontaneously generated' are *regular* kinds. However, it is not as if they present no problems: indeed some of those they pose are severe. The chief difficulty was this: while ordinary generation requires a species-specific formal and efficient cause (that provided by the male parent), how could generation occur *without* some such cause? If exceptions are allowed, do they not show that we need radically to revise the general assumption that in generation a formal and efficient cause of a very determinate type is required? So from that point of view, the fact that we have these regularly occurring instances of spontaneous generation raises more of a problem, for Aristotle, than one-off monsters or sports – for they can simply be attributed to a failure of the formal and efficient causes to do their normal work.

Similarly with cases of metamorphosis, for which Aristotle has no special term, though he describes the changes in the forms of

---

[2] Why there should be this difference in use need not concern us here. Balme 1962*b*, p. 91, suggested a developmental hypothesis (the *Metaphysics* antedating the zoological works) but it certainly does not seem necessary, in this case, to suppose that Aristotle changed his views, though he deploys the term very differently in different contexts.

certain insects clearly enough. How does this phenomenon square with the general requirement that a species should be defined in terms of its essential properties? Which, indeed, are the essential properties of a creature that at different stages in its life cycle, appeared to perform quite different vital functions – and not just as a matter of the regular acquisition of new capacities that was to be expected in the normal maturing of any animal as it approached adulthood?

So the two main questions we shall be concerned with are: first, what does Aristotle recognise as the data? What does he think has to be explained? Secondly, how does he explain them? How far does he recognise the challenge they present to his general theories and how does he meet that challenge? As usual, we shall endeavour, in the course of our study, to draw out some more general lessons that it suggests for Aristotle's interests, his theories and practices of science. We shall find, in this case, that he has a quicker eye for when the data he recognises can be used to defeat rivals than for the places where those data might threaten his own position. We shall also find that he acknowledges certain difficulties explicitly. They, however, in this case, do not lead him to make radical revisions of his key concepts, and he even manages to produce some arguments to turn some of them to advantage – though quite how satisfied he can be, or was, with the outcome is a tricky issue we shall confront at the end.

We shall approach this question by way of a comparison and a contrast with ideas formulated in a very different, but also highly sophisticated, ancient civilisation – with its own styles of learned inquiry – namely China. Learned or not, inquirers anywhere have well-known tendencies to see what they expect to. The advantage of the comparison with China is that it provides a test case from a culture whose expectations, on the questions concerned, were very different from those we can exemplify from ancient Greece. We need to focus here on China before the expansion of the influence of Buddhism, that is before about the +third century, since that introduced a set of exceptional factors promoting a belief in rebirth and metamorphosis.[3] But already before that expansion, in the Han period (roughly from the −2nd to the +2nd centuries) we

---

[3] See Needham 1956, p. 420.

have access to popular and learned texts that provide an insight both into ordinary beliefs about animals and into how animals fitted in to their world-views. What the Chinese saw, and what the Greeks did, are interestingly different – and we shall find another similar instance when we come to discuss what they saw in the heavens in chapter 8.

One particular text is especially useful to us as it offers some sustained comments on a variety of aspects of animal life in the context of a comprehensive cosmology. This is the *Huainanzi*, associated with prince Liu An.[4] The whole work was certainly not composed by a single author, let alone by Liu An himself, but it was put together under his direction around −139. Although its zoological sections, principally in *juan* ('chapter') 4, are far less extensive than the extant treatises on the subject in Aristotle, they cover a variety of topics, animal classification, methods of reproduction, periods of gestation, ecology, diet, pathology and 'evolution' (that is, origins and development).

It is important, first, to gain some idea of these general interests. Thus we find five main groups of animals recognised at 16A9ff., namely naked, feathered, hairy, scaly and shelled (or 'armoured'), exemplified by humans, birds, 'beasts' (*shou*), fish and turtles respectively. They are subsequently (*juan* 5) correlated with the five phases, *wu xing*, that is, earth, fire, metal, wood, water respectively. The phases are not elements in the Greek sense, since they are not fundamental substances. Rather they are modifications of *qi* (breath, 'pneuma') and they undergo constant change. Like everything else they manifest the interplay of *yin* and *yang*, correlative opposites that are defined in terms of one another, interdependent (*yin* does not exist without *yang*, nor *yang* without *yin*), functional and aspectual (the same item may be *yin* in one relation, *yang* in another).[5] As for the correlations between the five phases and animals, though proposed in *Huainanzi*, that idea was controversial. It was, for example, criticised by Wang Chong in the *Lun Heng* in the +first century, when he objected to its use to account for which animals prey on which.[6]

---

[4] *Huainanzi* chapters 3–5 have recently been the subject of an extended study by Major, 1993, on which I draw. On the strategic cosmological and moral interests of the work, see LeBlanc 1985.
[5] Cf. Sivin 1996.
[6] See especially the *wu shi* chapter of Wang Chong's *Lun Heng*.

Then a second more complex structuring appears in an admittedly very compressed form in *Huainanzi* 9B1ff. This deploys a number of paired differentiae, including (1) egg-producers and foetus-producers (i.e. ovipara and vivipara), (2) swimmers and flyers, (3) animals that swallow without chewing, and those that chew, that last opposition correlated with (4) animals with eight bodily openings and those with nine respectively. Further there are (5) horned and hornless, (6) 'fat' and 'non-fat', (7) with 'front teeth' (incisors) and without, (8) with 'back teeth' (molars) and without. These pairs combine in a variety of ways and yield not a single dichotomous hierarchy, but a complex polythetic network. In that regard *Huainanzi* would have sided with Aristotle in his criticisms of the dichotomists in *de Partibus Animalium* I 2–4.[7]

Again, like Aristotle, the interest in *Huainanzi* is not limited to physical differentiae, but includes also animals' habits, behaviour and characters. *Huainanzi* 8B1ff., for instance, remarks that animals that feed on grass are good at running but are stupid, those that feed on flesh are brave, daring, cruel, while those that feed on grain are knowledgeable, clever and short-lived. They are contrasted, in turn, with enlightened, long-lived sages who feed on *qi*, and with deathless sages who do not feed on anything at all.

The next section (9A3ff.) discusses the periods of gestation of 'humans, birds, beasts, the ten thousand creatures and tiny insects'. Humans are born in the tenth month of pregnancy, horses in the twelfth, dogs in the third, pigs in the fourth, apes in the fifth, deer in the sixth, for example. These perfectly credible figures are not merely factual and descriptive, since important numerological and cosmological correlations are also suggested. The seasons govern the pig, for instance, and so pigs are born in the fourth month: musical notes govern the ape, so apes are born in the fifth (five being the notes of the pentatonic scale).

We may turn now to the account offered of the origins and development of the five main kinds of animal, naked, feathered, hairy, scaly and shelled (16A9ff.). In each of the last four cases, the structure of the account follows the same pattern. We begin with a mythical creature, but one that has, as part of its name, the name of the kind that will come from it. Thus the feathered kind descend from Feathered Excellence (16A11), and the shelled from Shelled Pool. Those first creatures then produce a distinct type of dragon

---

[7] See especially *PA* 643b1ff., cf. above chapter 3 at p. 72.

Spontaneous generation and metamorphosis

(*long*) each, and passing through a number of other divine or fabulous creatures (they include the *feng* or phoenix, and the *qilin*), we come, in each case, to the ordinary or common (*shu*) birds, beasts, fish and turtles, from which are born the feathered, hairy, scaly and shelled kinds as a whole. The naked kind, and ordinary people, do not come via dragons, but via Oceanman and sages. But by 16B9 all five kinds are said to flourish in the outside world and to propagate according to type.

From a comparativist perspective, four features stand out. (1) There is no sense, in this account or elsewhere in the *Huainanzi*, of a *break* between divine or mythical creatures from past time, or creatures that have never been seen, and ordinary animals. From that point of view the account is more reminiscent of, say, Hesiod's *Theogony*, with its story of origins, starting with Chaos, proceeding through the genealogies of the gods and ending with mortal men and women. However the difference is that Hesiod does not talk of the origins of the different types of ordinary animals, whereas that is precisely what the *Huainanzi* is concerned to do. The common or garden animals we know all have origins that go back to the mythical creatures named.

(2) Secondly, there is no sense of the fixity of species, certainly not from cosmic beginnings. On the contrary, species of animals have a history and the origin of some groups can be traced to others. That is not to say that there is no sense of the stability of the principal existing groups themselves. They are clearly characterised with definite features, and, as noted, propagate according to type.

(3) Thirdly, a recurrent topic is their metamorphoses, not just their original transformations, but the metamorphoses they continue to undergo. Insects are not included in the account in 16A–B, but their metamorphoses are mentioned, for example at 9A11–B1. Nor are metamorphoses limited to insects, for at 9B3f. certain birds are said to change into clams. This was a theme that, as we have already remarked, was much developed in later Chinese thought under the influence of Buddhism. But *Huainanzi* antedates that influence and has no Buddhist morals to draw from the animal metamorphoses it refers to. But metamorphosis is not only common among animal kinds in the world of current experience: it also provides the articulating motif of the account of origins.

(4) Fourthly, *Huainanzi* operates without an overarching concept of nature. Indeed I would argue that such a concept is quite

absent from classical Chinese philosophy and science in general. By that I do not, of course, mean that they fail to describe and discuss many types of what we would call natural phenomena, using such concepts as *tian*, heaven, *tiandi*, heaven and earth as a whole, *wan wu*, the myriad things, *xing*, character – often applied to human natures – *li*, the patterns of things, *qi*, the breath manifest in the *wu xing*, the five phases, and so on. The term that forms the basis of the word eventually used in modern Chinese for 'nature', as in 'natural science', *da zi ran*, is particularly interesting from the point of view of our present concerns. *Zi ran* literally means 'self so' and corresponds to what happens spontaneously, that is without human intervention. In the chapter of the *Lun Heng* devoted to that very theme, Wang Chong's concern is to establish the limits, not of the natural as such, but rather of what is not intentional. He denies, for instance, that many phenomena that had been supposed to reflect purposes, or to convey signs, are, in fact, intentional at all. Prodigies are not warnings, nor are calamities sent to admonish humans.

But while classical Chinese philosophers and scientists bring to bear a battery of concepts in their explorations of how things are, there is no single overarching term that identifies the domain of nature as such, and marks it off from, on the one hand, the realm of the magical and the supernatural, and, on the other, what belongs to culture or convention, *nomos*. How those three domains were to be defined, and the relations between them, were disputed among the Greeks, not least in the context of controversies over morality: but that demarcations between them could be set up was one of the roles of the concept of *phusis*. Nor was this just a matter of defining things, but also people. Just as there was no single equivalent to *phusis* in ancient China, so we do not find any group of natural philosophers, *phusikoi*, there, intent on defining themselves by virtue of having that particular domain to investigate and by contrast with those who – in their view – ignored the boundary between the natural and the supernatural and confused the domain of the investigable with that of myth and magic.

If we may now attempt to summarise the outcome of this rapid glance at *Huainanzi*. It sees all living creatures as linked, and that includes sages, spirits and creatures of past times. As often elsewhere in ancient Chinese thought, there is no sense of fixity for ever, but an emphasis rather on process, interaction, interdependence. Cyclical change is, indeed, the key to the five-phase and *yin-yang* concepts, though we should be careful not to see those as already worked out in the archaic period. On the contrary, their systematisation was largely a product of Han times and does not long antedate *Huainanzi* itself.[8]

Animals are here used, in part, to locate humans by similarity and by contrast. As 'naked' creatures, humans are included in the five main classes of 16A9ff., just as they are mentioned alongside other animals when diet and feeding habits are discussed at 8B1ff., and when periods of gestation are in 9A3ff. But the spatial and temporal coordinates (as it were) of humans are given not just in relation to familiar animals, but also to the unfamiliar and to spirits. Ordinary people come from sages, we saw, and a further long section, 11B6ff., contrasts the civilised people of the central region with the inhabitants of the thirty-six countries beyond the seas, not just the Hairy people and the Hard-working people, but the one-legged people, the one-armed people, the three-bodied folk, people without anuses and the like. The patterns of similarities and contrasts here are complex. But while there is certainly a moral message here, it lies in the contrast between the civilised and the wild.

Neither here nor elsewhere is a notion of *nature* invoked to define the natural world, by contrast to the subject-matter of myth. The *logos/muthos* dichotomy is not applied, nor is it applicable. On the other hand, metamorphosis is, in this world, taken for granted in a great variety of past and on-going manifestations, and although the discussion of animals does not explicitly make the point, the same is also true of what happens spontaneously, *zi ran*.

Now due allowance has to be made for the differences in the scope and scale of the discussions in *Huainanzi* and in Aristotle. But what the *Huainanzi* focused on and what Aristotle did, are interestingly different. Of course one of the main articulating concepts of Aristotle's zoology – although it usually goes without saying – is that very notion of nature, *phusis*, itself, which we have just been discussing, and it structures his investigation in a number of ways. First, he is constantly on his guard, in the zoology as elsewhere,

---

[8] This point has been brought out forcefully by Sivin 1996.

against what savours of the mythical. He repeatedly criticises 'what is generally believed', 'what is said' or 'what is reported', when that seems to him unlikely or absurd – or he just suspends judgement and demands further investigation or verification. Particular writers, not just poets, but prose writers such as Ctesias and Herodotus, are rebuked for their gullibility. Thus Ctesias is said to be 'not trustworthy' (*HA* 606a8). When Herodotus is criticised for repeating a simple-minded but impossible story about the reproduction of fish (*GA* 756b5ff.) he is called a 'mythologist', *muthologos*. That term does not always carry pejorative undertones: Aristotle is, on occasion, careful to enlist the support of the ancients and their 'myths'.[9] But it can be used, as here, to imply a contrast with serious scientific inquiry.

Similarly Hesiod is sometimes spoken of in terms of some admiration, but he and the tradition Aristotle calls the 'theologists', *theologoi*, are often criticised for speaking unintelligible nonsense, as for instance at *Metaphysics* 1000a9ff., with regard to their ideas about principles and about the generations of gods and humans. Some of their views may be worth comparing with those of the Pre-Socratic philosophers (e.g. *Metaph.* 1071b26ff.), but they are not serious contributions to the detailed study of nature. However it is not that Aristotle is uncritical of the work of those he labels *phusikoi* either. When in his account of cosmology Empedocles imagined the generation of strange creatures, ox-headed humans and human-headed oxen, that gets dismissed as rank impossibility too.

What the inquiry into nature has to investigate is not such speculations, but, as we noted, what is true 'always or for the most part'. What is contrary to nature, *para phusin*, in the sense of the exception to the rule is recognised, but just as that, as exceptional, not as standing outside nature altogether.

Moreover the primary justification for the detailed study of animals, as Aristotle presents it,[10] is that this reveals their forms and final causes. So nature articulates Aristotle's account in this way

---

[9] Thus the lover of wisdom and the lover of myth are compared at *Metaphysics* 982b18ff. (cf. above, p. 104) on the grounds that myth is composed of the wonderful and that leads people to philosophise. Yet having the ancient myths on your side is one thing: depending on them to deliver the truth would be quite another. A classic case illustrating this dual attitude is *Metaphysics* 1074a38ff., the belief handed down that the heavenly region is divine and inhabited by gods. That is acceptable, though the accompanying anthropomorphic embroidery is rejected by Aristotle as 'mythical' (1074b4).

[10] *PA* I 1 and 5, cf. above chapter 2 at pp. 39f.

too, providing the answer to the question of what, to him, is valuable about the whole inquiry.

But if Aristotle's zoological agenda is the investigation of the orderly, indeed the fine and the beautiful, it is easy to see how this constrains his interests and throws up potential problems. The frequently reiterated principle, that 'a human begets a human', knocks on the head any idea of an evolutionary sequence. It is not that Hesiod or Empedocles before him, or Lucretius after him, had a theory of sustained evolution either, of course, but implicitly or explicitly they all denied the eternity of animal kinds and saw them as locked in a struggle for existence where only those that are strong survive. Aristotle, for his part, accepts struggle and explores it under the rubric of the *enmities* between animals in *HA* VIII and IX – exemplified by the relationship between predator and prey, for instance – the counterpart of their friendships. But the fixity of species is, for him, guaranteed by their breeding true to type. A monster is, by definition, a 'one-off' and does not reproduce itself.[11]

However the more Aristotle's zoology is driven by the conception of the norm provided by the dictum that 'a human begets a human', the harder it may appear to be to accommodate the real or assumed data encompassed by 'metamorphosis' and 'spontaneous generation'. Evidently in a world-view dominated by the Aristotelian conception of nature and natures, those two topics are likely to present greater difficulty than they do within the framework of classical Chinese conceptions. So we may ask how far Aristotle attempted to investigate them, what he was looking for or interested in when he did so, how far he recognised problems in what he found, and how he attempted to resolve them.

Considerable sections of *HA* V VI and *GA* III especially are devoted to the generation of a wide variety of kinds of animal that Aristotle believes to be produced 'spontaneously' in one or other of the senses already defined, chiefly that the animals have no parents, or at least not of the same kind as themselves. Apart from the 'spontaneous' (in this case, unfertilised) production of the eggs of fish and birds, he deals with other cases of spontaneous generation

---

[11] One should note that the anomalies allowed in Lucretius' account of the struggle for survival are strongly contrasted, by him, with the frightening beasts of Greco-Roman mythology. He has his own, but different, border to defend against the 'superstitious'.

of fish (e.g. *HA* VI 15) and more especially of insects and testacea. Practically all of the last group, he says, are produced without copulation (*HA* 546b15ff., cf. *GA* 762a8f.), and in *GA* III 11, especially, the spontaneous generation of testacea leads him into a discussion of the similarities, in this respect, between them and plants, to which we shall be returning.

Less space is given to metamorphosis, but both *HA* V 19 and *GA* III 9 especially discuss how a number of species of insects 'change their shape'. Thus at *HA* 551a13ff. we are told that the butterfly (the so-called *psuchē*) is generated from caterpillars which grow on the leaves of the cabbage and other plants. Initially it is smaller than a millet seed, but then it grows into a small grub and in a few days it is a small caterpillar. The transformations of some kind of silk-worm are described (551b9ff.), from grub to caterpillar to cocoon, though both text and interpretation are disputed.[12] The transition of the grubs of bees and wasps through a stage where they were known as *numphai* ('brides'), that is pupae, is also described, for example at *HA* 551a29ff.

A recurrent motif in these accounts is the difficulty of observation and the need for further research. Thus at *GA* 762a32ff. he remarks that of the testacea only the snails have been observed to copulate: but 'whether or not their generation is the result of the copulation has not yet been sufficiently observed'.

A more striking, because more detailed, case is the account of the *xulophoros*, 'wood-carrier', at *HA* 557b12ff.,[13] 'as strange an animal' as any he there reviews. First the head and the feet of the grub are described and how the rest of the body is closed in a cobweb-like membrane or tunic, surrounded by dry twigs that look as if they had become attached to it as it walks along. They are, however, an integral part of the membrane itself, and if one removes this, the animal dies, becoming as helpless as a snail without its shell. As time goes on, the grub becomes a chrysalis and is immobile. 'But which of the winged animals comes to be from it has not yet been observed' (557b24f.).

This instance illustrates both that Aristotle observed the effect

---

[12] It has been argued that this cannot be the Chinese silkworm on the grounds that this was not introduced into Europe until the sixth century. But Aristotle's text refers to some kind of fabric first woven by Pamphile of Cos. See, for example, Thompson's and Peck's notes.
[13] *xulophoros* is a plausible emendation for the MSS *xulophthoros*.

of certain interventions (the removal of the twig-like integument) and that he did not carry out a sustained investigation of the life cycle of the creatures in question, such as would have been possible if he had kept specimens under observation for a continuous period (for then he could have resolved the issue he raised at the end of the account). Whether or not he was himself responsible, he frequently alludes to what happens when certain animals are mutilated, or indeed cut up,[14] and this is not just a matter of idle curiosity, since the question of the maintenance or otherwise of the vital functions of an animal relates to the issue of its essence or defining characteristics.

On the question of spontaneous generation, too, we have signs of first-hand observation, but less of attempts to intervene to produce artificial conditions to test ideas. The need to give evidence and arguments for the data he accepts, let alone for his explanations for them, can be seen in many passages. Thus he is convinced that both a certain kind of mullet, and eels, are generated neither from eggs nor by copulation, on the basis of observation of dried out pools, HA 569a13ff., 570a3ff. What persuades him is that they have been known to reappear, after rain, in pools that had been completely drained of water and mud. And in the case of eels he is prepared to take issue with those who inferred that eels are generated from the small worms found inside them. They come rather from the so-called 'earth's guts' that are themselves spontaneously generated in mud: the eels have been seen (*ōmmenai*) freeing themselves from these, and they appear when these are pulled apart and dissected, HA 570a15ff., cf. GA 762b26ff.

The evidence of the dried-out pools is, of course, no artificially constructed trial to find out what happens, and, in general, experimentation on the conditions he believes accompany spontaneous generation does not occur to him, though he exploits exceptional circumstances for observation if he can.[15] It was enough, for his purposes, if it looked clear that the animals that came to be had not been produced by members of their own kind. Even the fact that he records in the case of the purpuras, HA 546b26ff., namely that they occur in greater numbers in places where their own kind

[14] For example, *de Juventute* 468b15, *de Respiratione* 471b19ff., 479a3ff., HA 519a27ff., IA 708b4ff., GA 774b31ff., and cf. further below chapter 7.
[15] Thus there are other cases where particular, exceptional observations are recorded, for instance at GA 763a30ff., 763b1ff.

had been before, does not arouse his suspicions.[16] Meanwhile while he certainly undertook many direct observations himself, there are many occasions when he was in some doubt as to whether or not to believe what had been reported to him by fishermen and others, and in many such cases he wisely reserves judgement.[17]

In much of the research he conducted to do with spontaneous generation and metamorphosis he clearly has a certain number of key questions in view. First he is concerned with the extent and nature of the vital faculties of the creatures in question. This corresponds, of course, to the cardinal importance of these factors in his definition of animals (cf. chapter 2) and in the distinction between them and plants (cf. chapter 3). Thus on the butterfly's transformations, he notes that as a caterpillar, it feeds and excretes, but as a chrysalis does neither (*HA* 551a24ff.). Again, while as a caterpillar, it moves, the *immobility* of the chrysalis is remarked, both here (*HA* 551a18) and in the account of the 'wood-carrier' (*HA* 557b23f.). There are not only surprises in the matter of changes in shape, but also in the loss of certain functions, normally thought to be acquired in regular sequence. At *GA* 758b28ff. he remarks on the problem, but dismisses the amazement that 'many' might feel with an explanation that appeals to an analogy with an egg: that too begins by growing, but then ceases to *grow* once it has become a perfect egg.

An even more fundamental question that preoccupies Aristotle with regard to many instances of spontaneous generation is the nature of the *causes* at work. Certain texts just deal with this in very general terms, the creatures are the result of the effects of heat – notably of the sun, or warm weather – on decaying matter. But some passages go into considerable detail on the precise kinds of matter in question.

Thus *HA* 550b32ff. discusses cases of spontaneous generation occurring from the following materials: dew on leaves, putrefying mud and dung, wood, hair, flesh, residues, whether excreted or not. Later texts add snow (552b6f.) and fire (b1off.) – surprising because seemingly such unpromising materials – and wool, clothes,

---

[16] This appears to be the point also made at *GA* 761b35, 762a2ff., though the latter text has been taken differently, for example by Peck.
[17] One such example has been noted above, chapter 3, on the question of the dispute over whether sponges perceive. I referred to several other cases in Lloyd 1979, p. 212.

cheese,[18] and books (557b1ff.). More important still, Aristotle associates particular kinds of matter with particular spontaneously generated creatures. Butterflies, for instance, come chiefly from cabbage (*HA* 551a13ff.), stag-beetles[19] from dry wood (b16ff.), ticks from dog's-tooth grass (552a15), horseflies from wood (a29), blister-beetles from the caterpillars on fig, pear or fir trees (a31ff.), gnats from the grubs produced in the slime of vinegar (552b4ff.) and animals like tail-less scorpions from books (557b8ff.). Elsewhere he discusses the differences in the different types of mud in which testacea grow and he goes into the varying marine habitats in which they are found in some detail (*HA* 547b18ff.).

We can see the importance of these moves. By assigning a determinate material cause to these instances of spontaneously generated animals, Aristotle brings them closer to normal patterns of coming-to-be. It is far from being the case that these creatures – or indeed any – come to be randomly in just *any* material. On the contrary, their matter has quite definite characteristics that correspond to their types.

The question of the material cause in spontaneous generation is, indeed, one of the two principal issues – the other being the efficient cause – that Aristotle explicitly recognises *as problems* when he comes to tackle the difficulties this kind of generation presents in *GA* III 11, 762a35ff., b4ff. In sexual reproduction the matter comes from the female parent who produces a species-specific residue from the 'useful' nourishment, and the efficient and formal causes are supplied by the male who produces his, more concocted, residue in the form of semen.

The first move to bring spontaneous generation into line with other kinds is to specify the matter – more or less determinately, as we have just seen. But the second is to see this generation too as the work of concoction (cf. chapter 4), and this in two respects. The first is that the efficient cause, for example the heat of the sun or warm weather, generally acts by concocting the matter (e.g. *GA* 762b14ff., cf. 743a35f.). The second is that the matter itself has, or at least may have, undergone a previous concoction, as notably in the case of creatures said to come to be in the residues – the dung,

---

[18] *kērōi* at *HA* 557b6 should no doubt be emended, whether to *pikeriōi*, with Peck, or to *turōi*, with Wimmer and others.
[19] Accepting the reading *karambioi*, with some MSS and editors.

for instance – produced by other animals. So although Aristotle frequently refers loosely to spontaneous generation occurring in putrefying matter, at *GA* 762a9ff., 13ff., he corrects himself and says that, more strictly, nothing comes to be by putrefying, but rather by a process of concoction, and the putrefied material is a residue of what has been concocted.

Thus far Aristotle has considerable success, by his own lights, in regularising the phenomena. But what about the formal cause in spontaneous generation, not taken up so directly in his review of the problems? From the point of view of the forms *produced*, there is no difficulty: the kinds of creatures spontaneously generated form regular species with definite characteristics, just as much as any other living being. True, some of them metamorphose and acquire and lose vital faculties in an initially puzzling way. But Aristotle does not remain puzzled for long on that score, and he is confident of the regularities of these metamorphoses just as he is of those of the creatures produced spontaneously.

However, the further difficulty relates to the absence, in cases of spontaneous generation, of a male parent pre-existing in actuality to provide the form. The heat of the sun can act perfectly well as an efficient cause of concoction, but cannot be considered to furnish a species-specific form.

Since he does not address that problem directly in those terms, the materials for his response – such as it is – have to be gleaned from two further points in the discussion in *GA* III 11. But before we turn to them, we should spell out a little more what is at stake.[20] At first sight the admission that a species-specific form is *not* always necessary seems to strike a body-blow at some of Aristotle's most cherished metaphysical convictions. If all that is needed, in some cases of generation, is the heat of the sun (say) acting on appropriate matter, why, we might ask, are male parents ever needed? Why, if matter does not need an independent form here, is one ever required? Is matter here made to do the work of form? Further, does this not threaten a collapse of the distinction between the living and the non-living?

The belief in cases of spontaneous generation, together with the usual doctrine that a pre-existing form-bearing male is normally necessary, makes, one would have thought, for a *very* difficult com-

---

[20] In what follows I have benefited greatly from incisive criticisms from Myles Burnyeat.

bination. Yet both those ideas occur, in close proximity, in *Metaphysics* Z9. At 1034b4ff. we are told that things come to be spontaneously when the matter can be moved by itself (*huph' hautēs*) with the movement that the seed (*sperma*) initiates.[21] Yet a little later, at 1034b16ff., he insists not just that in all cases of coming-to-be there is pre-existing form and pre-existing matter, but also that it is a peculiarity of substances that necessarily there must pre-exist another substance in actuality which does the making, and the example he gives is, precisely, of animal generation: 'for instance, an animal, if an animal comes to be'.[22]

Is it that Aristotle has been driven by his recognition of the facts of spontaneous generation, as he conceived them to be, into admitting a principle that offers a glaring exception to his usual doctrine of form, into granting, after all, that form is sometimes not necessary for generation? On that story, that would be a remarkable indication of his readiness to recognise empirical difficulties, but leave his metaphysics in serious disarray. He evidently never went back on the view that there *is* spontaneous generation in animals, as well as plants. And he also reasserts, in that text in the *Metaphysics*, the need for pre-existing form. How far can these views be reconciled? How far can his metaphysical doctrines be salvaged: or was he indeed himself in some doubt, if not confusion, on the question?

As remarked, there is no text that attempts a direct resolution of the problem we have posed. But two factors, both mentioned

---

[21] By that he might just mean with the *type* of movement that in *other* species the seed initiates. However, 1032a30ff. speaks of 'the same' animals coming to be both from seed and without seed, as may also be suggested at *GA* 761b23ff. Comparison with *GA* further shows that the capacity of the matter to move itself in the appropriate way is not the only part of the story, for spontaneous generation occurs under the influence of the extraneous heat supplied by the sun (e.g. *GA* 737a3ff.) or by the environment (e.g. *GA* 762b14ff.). However in *GA*, too, 762a18ff., as we shall see shortly, reference is made to soul-heat present in the material that produces animals and plants, so there too the possibility of an internal efficient cause is also envisaged. The clue that the *GA* provides for the understanding of this text in *Metaphysics* Z 9 is that it is earth or water endowed with a particular capacity that is at work in spontaneous generation. 'Matter', *hulē*, at *Metaph.* 1034b5, should, then, be interpreted as 'stuff': see below, p. 123. Some of the opacity of the argument in *Metaph.* Z 9 may also stem from Aristotle's bid to assimilate the spontaneous generation of animals to the production of health, which may happen by the doctor's art, or spontaneously, but in both cases is brought about by heat (*Metaph.* 1034a9ff., 26ff., carrying on the topic broached already in Z 7, 1032a28ff., b4ff., 21ff.).

[22] Further puzzles raised by this highly problematic passage, for instance what he means by saying that in the coming-to-be of quantities something that potentially has the quantity in question pre-exists, need not, fortunately, concern us here.

in the key chapter, *GA* III 11, deserve particular attention. First, Aristotle has a highly hierarchical and normative notion of nature and stresses at *GA* 762a24ff. that there are differences in the value, *timē*, in what is formed. Secondly, he is evidently impressed by the creative potentiality or productivity of nature.

Both points need some elaboration. Most of the animals that come to be spontaneously and that undergo metamorphosis are low on Aristotle's scale of being. But as he goes down that scale, Aristotle tolerates all sorts of phenomena that he is prepared to label 'deformities' (*anapēron, pepērōmenon*), even though they are regular, not monstrous. They are 'deformities', that is, judged by the norms provided by higher animals, especially humans, indeed particularly male humans since females, as he often says, are like 'natural deformities', in their inability to concoct semen (cf. chapter 4). But he is prepared to say, for example, that judged by humans all other animals are 'dwarf-like', in that they have their 'upper' parts, that is those near their head, larger than the lower ones.[23]

We shall have more to say on this idea of degradation later, but for now may note that the fact that the testacea bear many points of resemblance to plants is here not a *problem* (to do with the boundary between animals and plants, as we discussed in chapter 3), but can be turned to advantage. Many plants, to be sure, produce seed and are, in that, like animals that do so, even though in plants, male and female, according to Aristotle, are not differentiated, but, in a way, combined (*GA* 762b9ff.). But spontaneously generated plants come from the action of heat on the earth, and in that, provide a close analogue to spontaneously generated animals in the earth and the sea.

At *GA* 762a18ff. he has this remarkable statement to make about what I called the creative potential of nature. 'Animals and plants are formed in the earth and in the moist, because in earth, water is present, and in water, *pneuma* is present, and in all *pneuma*, soul heat (*thermotēs psuchikē*) is present. So that in a way (*tropon tina*) all things are full of *psuchē* (life/soul).'[24] So the absence of a species-

---

[23] See, for example, *PA* 686b2ff., b20ff., 689b25ff., 695a8ff. The topic of such comparative natural deformities is discussed in some detail in my 1983, pp. 40ff.

[24] 'All things' here should, presumably, be taken loosely as referring not to every single type of substance – rocks, say, or diamonds included – but just to all the types of material in which plants and animals are generated. However, the list of these is very considerable, as we have seen, pp. 116f.

specific form here turns into an argument for the particular productive capacity of nature working through the *pneuma* instinct with life.

That absence is, no doubt, a mark of the lowliness of the living beings in question. From some points of view, they could be said to resemble the homoeomerous substances, metals or stones, produced by combinations of earth and water that Aristotle discusses in the *Meteorologica*, for they too are not the result of the imposition of forms provided by actually pre-existing species-specific individuals. However, they are *in*animate, and so yet further below the plants and animals in question in *GA* III 11.

A closer analogue would be the spontaneously produced eggs of fish and birds we mentioned before (pp. 113f.). The hen lays infertile eggs without the intervention of the cock. So thus far an external male efficient cause is not required for this coming-to-be. Yet this is not a perfect analogy either, not just because the eggs are infertile, but also because the hen herself is, of course, a member of the species in question, even though – as Aristotle would no doubt insist on adding – a defective member.

Yet while the creative potential of nature would supply some part of an answer to what happens in the case of spontaneous generation, may it not raise more difficulties than it resolves? The presence of soul-heat in all *pneuma*, and the view that 'in a way' all things are full of soul, are immediately suggestive, of course, of hylozoism, the idea that matter itself is instinct with life, a view which, when he associates it with Thales' use of the principle, water, in *de Anima* 411a7ff., is promptly criticised on the grounds that it provides no explanation of why there should be living creatures in some water but not in other water.

Yet perhaps Aristotle has a defence to *this* objection, one that would again invoke nature and natures. He was, he could say, not using his idea of the role of soul-heat in producing animals in any random fashion, but rather working back from given explananda, namely the facts of the generation of certain specific plants and animals in certain conditions in certain materials, conditions and materials that he had specified, at least up to a point. As to why nature did not elsewhere produce other creatures, no explanation could be given for non-events. Yet one was clearly needed for those spontaneously generated animals and plants he had identified, even though that explanation was not in a position to offer independent specification of the limits of its applicability – independent, that is, of the very explananda it was invoked to deal with. But so far from being committed to hylozoism in general, all he was committed to was the operation of soul-heat in *pneuma* in these specific cases, to explain their specific natures. Whether Thales himself might have had a similar line of defence to the objection Aristotle brings against him is a question we shall never be in a position to answer. But despite the apparent generality of 'all things are full of soul', Aristotle allows himself only a limited appeal to the soul-heat in *pneuma*.

So how successful is he, in the final analysis, in salvaging his metaphysical principles in the face of such data as he recognises? In metamorphosis the problem can be said to be a matter of deciding at which stage in the life-cycle of the creatures in question we have the mature adult. But that is primarily an empirical question, and once it is resolved, we have, potentially, the answer to the question of the essence of the species.

Spontaneous generation is, undoubtedly, more problematic. At least we can see that there is *a* role for all four causes. First there is matter: indeed in many cases, as we saw, he insists that the matter in question is quite determinate, different types of mud, or wood, or even dew or snow. Secondly, there is the efficient cause. This is represented, on different occasions, as the sun, or the warmth of the environment, or just as the soul-heat in the *pneuma* in the water, but it is responsible for achieving a due proportion, *summetria*, of heat for the matter concerned. Thirdly, there are final causes – the creatures produced, which, insofar as they form regular species of animals, as they do, also represent, fourthly, the formal causes.

But the difference, or anomaly, lies, of course, in the way these causes work. In humans, say, the form goes with both the final and the efficient causal factors and is contrasted with the material, whereas in spontaneous generation, form is still associated with final, but *not* with the efficient, causes. The forms can still be said to pre-exist – indeed they are eternal – for Aristotle no doubt believed that there have *always* been the creatures in question. But the pre-existing form does not also serve as the efficient cause (as male parents do). In spontaneous generation, the material in question has to take on a more active role. Some stuff – mud, slime or whatever – has to be thought of as endowed with soul-heat and is even able to be moved 'by itself', as in the *Metaphysics* text, 1034b4ff.

But those admissions do not mean the end of the distinctions (1) between the living and the non-living, and (2) between the matter and the formal and efficient causes, even though a delicate balancing act has to be performed, since the more the material is treated as active, the more hylozoism threatens. But as to the latter distinction, the defence could be that it is not matter *qua* matter that serves as its own efficient cause, rather that certain material has a certain extra capacity: we can remind ourselves that we never encounter matter *as such* in nature, but always stuff of a certain kind. Then as to the former distinction, where we have already seen that Aristotle problematises the boundaries (chapter 3), we can still see that Aristotle can and will insist that the mud is not itself alive – it does not fulfil all the criteria for 'life'. It is still very different from plants, let alone animals, even if it does have 'soul-heat' that, under certain conditions, produces various types of animals.

Aristotle has to make certain concessions, in the face of the exceptional circumstances of spontaneous generation, but they are, I would argue, in line with his general views on the graded hierarchies of nature (a recurrent theme in *GA* III 11).[25] He is clear that the norm is provided by the higher creatures, pre-eminently humans. In them, male and female are distinct and given separate roles.[26] But in some animals, and in plants, male and female are not distinct. There, the female has the male principle 'mixed in' (*GA* 762b10f.) in the living beings themselves. Spontaneous generation represents a further departure from the ideal. The fact that mud can produce testacea can be seen as a further, indeed extreme case, of the non-differentiation of roles: for the heat in this case – whether the soul-heat in the mud, or the heat supplied

---

[25] *GA* 762a24ff., mentioned above (p. 120) is not the only example, since he begins the chapter with some elaborate speculations correlating different types of creature with the places or elements where they live, *GA* 761a13ff., and especially b13–22.

[26] A Cambridge doctoral dissertation by Sophia Elliott, in progress, will, however, show that it is misguided to underestimate the positive roles assigned to the female parent, not least in the transmission of certain of her characteristics to her offspring (*GA* IV 1–3).

by the sun or the warmth of the environment, or of some combination of those – is entirely *unspecific* to the creatures in question.[27] That does not make the creatures superior, of course: quite the reverse. For Aristotle, in this as in other domains, it is better for high and low to be distinguished and separate, whether male and female, or master and slave.[28] His view of the matter is that that allows the higher to fulfil *its* superior functions to the full, even if there is a price to pay for this, namely that the lower is limited to *its* inferior role.

One thing that emerges from this study of Aristotle's treatment of spontaneous generation and metamorphosis is the robustness of his views on nature. Where the assumptions the classical Chinese commentators on similar phenomena worked with stress interaction, change, what we may even call the plasticity of things, and where they recognised no sharp boundaries between what the Greeks contrasted as the natural and the mythical, Aristotle uses his notion of *phusis* both destructively and constructively.

Destructively, it is the cornerstone of his refutation of his rivals whether from among the 'theologians' or from other would-be 'natural philosophers'. Both Hesiod and Empedocles, for example, were guilty of wild speculation that was far above everyone's heads: and neither appreciated the regularities of nature.

Constructively, nature, corresponding to what happens 'always or for the most part', both identified the explananda and provided a framework for their explanation. Thus what might *seem* strange or puzzling phenomena, spontaneous generation and metamorphosis, are brought within its orbit and thus far regularised. The puzzles can be resolved at least insofar as the species in question have the determinate characteristics they have, and undergo the transformations they undergo, in a regular fashion. In the case of spontaneously generated creatures, they can be given material causes, indeed in some instances quite determinate ones, and the efficient causes too can be described. While both spontaneous generation and metamorphosis depart, in certain respects, from the patterns set by higher animals, the hierarchy of the scale of being

---

[27] In the case of the soul-heat in mud, there is a further lack of differentiation, in that the efficient cause is internal, not external, but then that is often, indeed, generally the case in nature.
[28] Cf. below ch. 9, pp. 189ff.

allows for, and even insists on, a gradual decline, from those animals, to lower ones, to plants and on to inanimate objects.

It cannot be said that Aristotle shirks the problems: he faces them with remarkable frankness. Nor does he attempt to dismiss the phenomena that seemed to cause difficulty, and he points out that much needed research remained to be done. Yet this is an instance where his leading idea, the concept of nature itself, is not so much modified, let alone undermined, as confirmed. True, that concept is not his single-handed invention, though it certainly underwent elaboration and systematisation in his hands. But compared with the problems he still had, those of his rivals undoubtedly seemed far greater when they failed to acknowledge the regularities of nature.

It is thanks to his research into such regularities that the concept was not merely elaborated but confirmed. Its ability to encompass what seemed anomalous phenomena is turned into one of its strengths. What *remains* wonderful is not what the mythologists thought, but nature itself. So the wonderful does indeed remain the prompt to philosophise – provided you understand what is truly wonderful.

The comparison with classical China serves to show how much more open-ended ideas about what is possible could be, and were, in the absence of the idea of nature. We shall see other occasions later where the corresponding open-mindedness enabled the Chinese to see what Greek ideas of an unchanging universe ruled out (see below chapter 8). Its power as an organising principle in zoology, however, is clearly exemplified in Aristotle's work on spontaneous generation and metamorphosis. It is his notion of nature, above all, that he uses not just to delineate the boundaries within the subject-matter he investigates, but also to police those between his kind of investigation (natural philosophy) and those of other fellow-Greeks he is only too eager to downgrade or to dismiss.

CHAPTER 6

The varieties of perception

Aristotle's doctrine of perception plays a key role at the intersection of a number of his deepest concerns. It is fundamental, naturally, in his account of cognition and in his own philosophical and scientific methodology; it provides the main distinguishing mark that differentiates animals from other living beings (cf. chapter 3); in ethics, it is basic to his account of practical action, being responsible for judgement concerning particulars (for example, *Nicomachean Ethics* 1126b3ff.).

The central features of the general account of perception are well known. It is the reception, by the sense organs, of the perceptible forms without the matter (e.g. *de An.* 424a17ff.). Each of the senses has its own proper objects, the perceptible forms, analysed as pairs of opposites, such as white – or light – and black – or dark – between which the sense-organ is a mean, liable to injury, in the case of touch indeed leading to the death of the animal, if affected by extremes. But while each of the individual sense-organs has its part to play, perception as such is the work of the common sensorium, located in the heart (or the analogue).

The terms in which the general theory is set out are agreed. Yet their interpretation remains, at points, the topic of intense debate. One such controversy concerns the reception of the perceptible form. When the eye sees a red object, does the eye-jelly itself become red? When the ear hears a sound, does it itself resound? There are those who agree with Sorabji in answering yes to both questions, and others who, following Burnyeat especially, deny that that is right. Perception is not a matter of being qualitatively affected, so much as, precisely, one of discrimination, *krinein*. We become aware of the perceptible forms but are not qualitatively changed by them.[1]

[1] Apart from Sorabji 1971/1979, 1974/1979, 1991 and 1992, Burnyeat 1992 and 1993, see

There is a certain indeterminacy in the Aristotelian texts that can be cited on either side of this dispute, which is, in any event, a dispute not just about texts but also about the modes of explanation that Aristotle attempted or that can be expected from him. My aim, in this chapter, is not so much to contribute to that debate, as to tackle a group of questions to do with the *generalisability* of the model of perception that Aristotle proposes.

It seems clear that he *seeks* a general model applicable to all five senses. But how well does he succeed in that aim? How does he deal with the variations that he evidently recognises between the five senses, especially those between touch on the one side, sight, hearing and smell on the other? Again, extending the range of the discussion to include animals other than humans, I wish to ask how far his general theories are able to explain the varieties of animal perception. The fundamental mode of perception is touch, mediated through flesh, and Aristotle has a carefully worked out account of the different modalities of flesh in different kinds of blooded and bloodless animals, relating, for example, to whether the flesh is *inside* or *outside* their bony structure. Can Aristotle's theory of perception account for these variations too? Does he recognise the problems and if so, how does he respond to them? In what will again be a speculative discussion, I shall offer some suggestions that will attempt to throw further light on the extent to which he seems prepared to modify and adapt his general theories in the face of difficulties.

Our first problem is the relation between his account of touch and that of such other senses as sight and hearing. In a number of obvious and fundamental ways, the theory of touch parallels that of the other senses perfectly. It too is the reception of perceptible forms, it too is a mean between opposite extremes, it too is, strictly speaking, a function of the common sensorium, happening 'inside', namely at the heart, rather than at the surface of the body.

We may note, however, first that the perceptible forms are more

especially Ebert 1983, Modrak 1987, Lear 1988, pp. 101ff., Freeland 1992 (Waterlow) Broadie 1993. The further debate, mentioned above (ch. 2 at p. 62), between functionalists and their opponents, is connected with the Sorabji–Burnyeat controversy at least insofar as the functionalists presuppose a Sorabji view on the role of an affective change in perception. In his latest article, 1992, Sorabji himself has modified his view, as far as functionalism is concerned, though not on the eye-jelly going red, when seeing red, or more generally on the role of a qualitative change as material cause in perception.

varied than is the case with sight or hearing. It deals not just with one primary pair of opposites, but with several, hot/cold, wet/dry, hard/soft and so on (for example *de Anima* 422b25ff., 423b26ff., *PA* 647a16ff.). Moreover where, in the other senses, when an extreme affection occurs, it is just the sense-organ that is destroyed, in the case of touch, as noted, the animal itself will die (*de Anima* 435b13ff.).

As for touch connecting with the heart, surprisingly, perhaps, he is more confident that this is the case than he is with the other senses. They of course were located in the head, and he is aware of certain connections, *poroi*, between the eye and the ear, for instance, and the back of the brain (for example *PA* 656b16ff., b18f., cf. Lloyd 1991, pp. 237ff.). He even has arguments why it is reasonable that sight should be near the brain – since the brain is wet and cold, and the sense-organ of sight is wet (*PA* 656a37). Yet it is obvious, he claims, that touch and taste extend to the heart (*de Juventute* 469a12f., a20f., *PA* 656a29ff.), and in this case he extrapolates from them to the others: 'so the others necessarily do too' (*de Juventute* 469a14).

What precisely he has in mind in the case of the connections of touch and taste with the heart is unclear. But we must remember that flesh is the medium or organ of touch, and that just as both the bones and the blood-vessels each form a single, continuous structure through the body, so the flesh *is* in a sense the body (*PA* 653b21f., 654a32ff.) and certainly communicates with the heart via the blood-vessels. That leaves the question of why touch should be thought to be connected specially to the heart – rather than to any other part of the body – but the answer to that lies, of course, in the independent grounds he would claim to have for his doctrine of the primary role of the heart as the control centre of the vital activities of the animal.

The first important potential disanalogy between touch, on the one hand, and sight and hearing, on the other, is one that bears on the controversy mentioned at the beginning of this chapter. We may begin by remarking that the perceptible forms in the case of touch differ from those of sight and hearing in this respect, that the flesh itself, like any physical object, is characterised by the very qualities, hot, cold, wet and dry, that it has the task of perceiving. Whether seeing involves the eye-jelly going red is disputed. But when we touch something hot, heating of some kind is undoubtedly involved. Aristotle says as much in *PA* II 2, 649b2ff., cf. 648b12ff., b30ff., 649a10f. – all texts where he refers to heating in respect of touch. True, that is the chapter we have mentioned before where he is very careful to insist that hot is 'said in many ways'. So the question of *what kind* of heating is involved when something hot is touched is an open one. We can distinguish, say, the effect of boiling water – which is said to be hotter to the touch than oil – from the burning and melting caused by flame, 648b26ff., 30ff. Yet that some heating takes place is clear, and on the view I would favour that is the factor that we need to account both for the fact that we do not perceive what has the same temperature as ourselves (*de An.* 424a2f.: in such cases there is no heating or cooling) and the further fact that when exposed to heat or cold in excess, the sense-organ is destroyed and indeed with it, the animal itself (*de An.* 424a28ff., 435b13ff.).

At first sight this might appear to support the view that has it that in vision, too, the eye-jelly goes red. However, that is to jump to conclusions. First – the issue we are interested in here – the question of whether analogies hold between the senses cannot be prejudged. We are concerned to explore whether or to what extent Aristotle has a *general* theory and what strains generalisation is under.

Secondly, and more directly to the point, even if the flesh is heated, whether the heating *is* the perceiving is entirely another matter. No doubt the heating may be considered a necessary condition for the perceiving. But that it is *not*, in itself, the perceiving is strongly suggested, for instance, by the argument from sleep. When we are asleep, our body, our flesh included, undergoes changes in temperature which we do not perceive. Sleep is indeed caused by such a change, though an internal one, the retreat 'inside' of the heat (*Somn.* 457b1). So since these qualitative changes occur, at the common sensorium indeed, without being perceived as such, they cannot be identical with the perceiving. The question of what the 'reception of the perceptible form' is remains open, but it is clear that the case for considering it *essentially* a discriminating is strengthened, not weakened, by the case of touch.[2]

---

[2] Sorabji would, no doubt, reply that the touch case shows that there the qualitative change is (at least) a necessary condition for the perception. Yet I would insist first that the heating is not constitutive of the perception, and secondly that in any event we

But a second potential disanalogy is, for our purposes, more important still. Is flesh the sense-organ or the medium of touch? Aristotle himself hesitates on the question, and this is no mere residual doubt on a peripheral matter, after all the main points of doctrine have been settled. Why does he hesitate and what are the implications for his general theory?

On the one hand he refers readily enough, in no doubt loose terms, to the flesh as the sense-organ, *aisthētērion*, of touch. He does so, for instance, at PA 647a19f. On the other, at PA 653b24–30 and again at *de Anima* 422b34ff. he mentions both possibilities. In the first of these contexts he is concerned with the role of flesh, and in the second focuses on what the sense-organ of touch is, and in both he eventually concludes in favour of the view that flesh serves (at least in part) as the medium of touch.

Thus at PA 653b24ff. his preferred conclusion is that flesh is not so much the primary sense-organ (the analogue of the pupil) as that taken in combination with the medium (as if one were to add the whole of the transparent medium to the pupil). That still leaves flesh with a role as part of the sense-organ apparatus, and it obviously could not serve just as an *external* medium, since that would leave quite unexplained why this or that animal perceived this or that hot or cold object.

His most extended discussion of the problem comes in the second passage cited above, *de Anima* II 11. There he undertakes to show first that in the case of touch, perception takes place 'inside', not on the surface of the body, and secondly that, despite appearances, touch requires a medium just as much as the other senses do. Both conclusions serve to bring touch into line with the general theory, but leave certain differences between the accounts of the five senses still in place.

Thus at *de Anima* 422b34ff. he asks where perception takes place: is the sense-organ inside, or not? Does the flesh itself act immediately as such? The fact that we perceive as soon as we are touched is no indication. Even if we stretched a membrane over the flesh, the same experience would occur: but clearly the membrane is not the sense-organ.

cannot extrapolate from the touch case to other senses for the reasons I have given, namely that it is a distinctive feature of touch that the primary tangible qualities belong to any physical object and so also to the sense-organ in this case.

That thought experiment is used again at *de Anima* 423b6ff. to show that a *medium* is necessary in touch. Even if we perceived all the 'tangibles' through a membrane, we would not notice this: for that would be the same as now happens to us in water and in air, where, as it is, we do not perceive the medium that surrounds us.

However, there is still this difference. In sight and sound we perceive thanks to the medium 'doing something' to us, but in the case of touch we are not affected *by* the medium, so much as *together with* the medium, 'just as if someone were struck through a shield' (423b15). 'It is not that the shield, having been struck, knocked the man: but rather what happens is that both are struck together.'

This enables Aristotle to suggest that in *no* case does perception take place if the sense-organ as such is in direct contact with the perceptibles in question. That is obvious in the case of sight and hearing and smelling. But the same is taken to apply in the case of touch as well, and the fact that we perceive when tangibles are in contact with the flesh is taken to indicate that the flesh is the *medium* – rather than the sense-organ – of what has the faculty of touch (423b23–6, cf. 426b15ff.).

We shall come back to the thought experiments with the membrane and the shield in a moment, but first we must take stock of the extent to which Aristotle has, thus far, secured uniformity across the five senses, and the price he has paid for this.

First he shows no doubts whatsoever on the question of the common sensorium. All modes of perception, without exception, are the work of the common sensorium located 'inside', that is at the heart. At this point there is no problem in having touch (and taste) coming into line and they might even be said to make the running in the argument for their 'obvious' connections with the heart.

But then on the question of the need for a medium, there is some initial hesitation in the case of touch. 'Touching' is a matter of 'contact' – as indeed the Greek term *haphē*, used for both, suggested. On that score there seems no call to treat the flesh as anything other than the sense-organ. Yet we have seen that that is not Aristotle's preferred position, which has the flesh serve as the medium of touch, even while preserving a difference between perceiving 'at a distance' (as in the case of sight and sound) and perceiving 'close to' (as in the case of touch, *de An.* 423b2–6).

At this point it might be suggested that the needs of his general theory do influence Aristotle to the extent that he relies on analogies to support the conclusion that in touch, too, a medium is necessary. But there are further factors as well that point in the same direction. So far we have been talking mainly about Aristotle's account of human perception, though such expressions as 'flesh or its analogue' remind us that he has perception in animals as a whole in view. But now I want to pursue the question of how his account of touch fits what he has to say about other animals.

Touch is the essential mode of perception of all animals, *qua* animals. But how does Aristotle deal with the apparent difficulties that might be thought to arise in the case of those animals who not only do not have flesh, but just its analogue (the 'fleshy', *to sarkōdes*) but also have that 'fleshy' part '*inside*' rather than on the outside of their bodies? The argument will be that – whether or not devised in part for this purpose – the analysis of *de Anima* II 11 provides an essential ingredient in the solution to the problems posed by the sense of touch in some of the lower animals.

It is the bloodless animals that cause the most difficult cases. The Cephalopods can be accommodated fairly straightforwardly, for they are more or less completely fleshy and soft (*PA* 654a12ff.). The problem that arises with them is that of securing a strong enough structure for their bodies without bones as such. The way that is resolved is, first, that their flesh is midway between flesh and sinew, and secondly that some of them do have something analogous to fish-spine in fish, namely the 'pounce' of the cuttle-fish and the 'pen' of the calamaries.[3]

For the insects, the account offered of their structure similarly leaves open, implicitly, the possibilities for their possessing touch. At *PA* 654a26ff. Aristotle says that with them there is no clear-cut distinction between hard and soft, and the whole of their body is hard: yet its hardness is more fleshy than bone, and more bony and earthy than flesh (cf. also *HA* 523b15f., 532a31ff.). That ensures that their body is not easily cut up, but at the same time Aristotle would no doubt claim that the medium of touch is provided, for them, by the comparative fleshiness of their bodies.

With the testacea, however, the problem is more acute, and acutest of all with the crustacea. Both these groups have the hard

---

[3] Yet as *PA* 654a22ff. recognises, no equivalent structure exists in the octopuses, where the 'head' forms only a small sac.

supporting structure outside, and the fleshy part inside. That arrangement serves a variety of purposes, for instance it is said to preserve the little vital heat that these creatures have, *PA* 654a5ff. But on the face of it, to have the bony structure outside might well be thought to constitute a severe impediment to the sense of touch. At *de Anima* 435a24ff. Aristotle states that we do not perceive in our bones, in our hair and suchlike parts, because they are earthy, and that is there given as the reason why plants do not have any perception, since they are 'of earth' (though elsewhere their non-perceiving is put down, rather, to their lacking a mean, *de An.* 424b1.)[4]

But even if an animal cannot perceive *with* bone, or with the bony, the crustacea and testacea clearly have to perceive *through* their bony structures, and that implies firstly that the tangible forms are transmitted *through* the bone and that – that done – they are then transmitted on to the common sensorium in the analogue of the heart. We shall discuss the first of these questions shortly, but the answer to the second is clear: what serves the second function is their 'fleshy' part.

So here we have a case of touching where the medium itself is *inside* the body. That might look like an exception to the general observation, made in the *de Anima* 423b3, that touch operates by contact (for that appears to leave no room for indirect contact), but it is no exception to the rule that flesh is the medium and that it is when the tangible forms affect it that touch occurs.

Indeed we can see that the crustacea and testacea cases are precisely analogous to the thought experiment situations, explored in *de Anima* II 11, in which we perceive *through* a membrane or *through* a shield. We cannot say that those thought experiments were directed to solving the crustacea and testacea problem: that is not mentioned at this point in this chapter, even though in the opening discussion, 422b20f., reference is made to 'flesh or the analogue'. The thought experiments play their own parts, as we have seen, in their own arguments concerning the seat of perception being inside and flesh serving as the medium. Yet what they offer is support for the possibility of touching *through* an interven-

---

[4] Yet as *de Anima* 424a32ff. acknowledges, tangibles do have an effect on plants, for they are heated and cooled – another text that demonstrates what in the Sorabji–Burnyeat controversy is not, in any case, in doubt, namely that heating and cooling are not themselves sufficient conditions for perception.

ing solid object, and that is precisely where the difficulty may be thought to arise in the crustacea and testacea cases, since they are surrounded by their non-perceptive – bony structures.

The shield case, especially, shows that touch can occur even when the intervening object is substantial. Nor would it be difficult to explain why Aristotle would prefer to cite the shield case, rather than the crustacea, at this point in his argument, since, first, we can verify that he is right from our own sensory experience, and secondly, he does not in any case as a general rule deploy lower animals to explain higher ones, but rather quite the reverse.

Yet such comments as these do not meet all the difficulties. Consider *which* tangible properties are transmittable, through a membrane, a shield or a bony structure. Impact will be (though that is not a tangible contrariety) and so will, up to a point, the hot and the cold. Even there, however, there is a problem in that (as noted) one of the functions of the bony structures of the crustacea and testacea is to preserve their vital heat. Since their bony structures are certainly not represented as *one-way* permeable membranes, they are presumably a poor medium not just for letting (vital) heat out, but for allowing (tangible) heat in.

But what about the wet and the dry, and the other differences that touch discriminates, including hard and soft themselves? The softness of the seaweed brushing against the shell of an oyster is not easily going to be registered inside that shell (yet it needs to be aware of it in one way or another, since it is said to feed on seaweed, *HA* 568a4–8). Even if the task of discriminating the wet and the dry will hardly arise much with those creatures that spend their whole time in the sea, not all the crustacea do (the crabs have a variety of habitats, e.g. *PA* 684a1ff.): nor do all the testacea (for example the land snails, *HA* 528a8f.).

Since Aristotle does not deal with *these* problems directly, his response can only be guessed. One line of argument might be that his account aimed to secure touch as such for these creatures, not necessarily the perception of the whole gamut of the tangible forms, and a second might be that the chief mode of touch that is essential for the lower animals is taste and that is mediated through the perceptible apparatus in the mouth of the animal, especially its tongue.

Yet, to the second argument first, though there are many occasions when taste is assimilated to or associated with touch, they are still distinct modes of perception. It is always touch that is said to be the primary mode, that is often asserted to belong to all animals (*de An.* 413b4f., 414a2ff., b3: but cf. chapter 3), indeed to be the only one to do so (*HA* 489a17f., *Somn.* 455a27). So we are owed an account of touch and not just of taste – not least, indeed, since in the case of the crustacea he says that they have all the senses (*HA* 534b15ff.) and in that of the testacea, he explicitly distinguishes their sense of taste (and smell) from that of touch (*HA* 535a4ff.).

The first line of argument, if weaker, may be more likely. Aristotle is recurrently concerned with the excellence, or otherwise, of the various senses in various kinds of animals, where he often has the comparison with human perception, in particular, in mind. But by the same token that allows him to attribute a weak, indeed a minimal, capacity to some animals with respect to some senses. It is not always the case, though it often is, that humans provide the model of the best endowed animal (cf. Lloyd 1983, pp. 26ff.). At *Sens.* 440b31ff., indeed, humans are said to have the worst sense of smell of all the animals, and *de Anima* 421a20f. says we fall behind in others, too. On the other hand at *HA* 494b16–18, at *PA* 660a11ff. and at *de Anima* 421a16–20 he makes a particular point of the claim that humans have the most accurate sense of touch – and so have the softest flesh.

But if Aristotle does not show an interest in meeting the detailed objection I have raised about the sense of touch in some of the bloodless animals, his overall concern with the *general* thesis that every kind of creature has flesh or its analogue is not in doubt. That remains fundamental in his account of animals qua animals – precisely because of the role of the 'fleshy' in the basic perception, touch. It occupies his attention in his account of each of the bloodless kinds, where, as we have seen, the varieties of their 'fleshy' parts are gone into with some care. It is all the more striking, therefore, that – as we also found in his treatment of the possession of perception itself, chapter 3 – he is prepared to allow exceptions to what might have been imagined to be inviolable rules. At *PA* 679b34 one of the testacea, the sea-urchin, *echinos*, is said to have *no* fleshy part,[5] and at *HA* 528a6f., too, it is cited to show that that is true of some of this group of bloodless animals.

---

[5] That is, not even an analogue for flesh, since that is what he is talking about in the context.

In this study, as in several of our others, we find that as soon as we press the details of the account Aristotle offers, a tension is revealed between what his general theories suggest or require and what his appreciation of the complexity of the data leads him to acknowledge. The first point that is worth emphasising, in conclusion, is just the quality of that appreciation. This is no armchair commentator satisfied just with the grand generalisation – on the cognitive capacities of the animal kingdom or on any other issue. It is his very concern with the fine details of the account that on the one hand gives it such authority, and on the other brings about that tension.

On the question of perception, the bid to develop an overall theory answers to the requirements of his view of the role that perception as such plays in so many areas of inquiry. Yet faced with divergences first between the five senses in humans, then between the modes of perception in different kinds of animals, he does not smooth them *all* out, even though some of the explanatory moves he makes have that tendency.

Thus his preferred resolution of a certain doubt about the role of flesh in touch – is it the sense-organ or the medium? – is that it serves the latter function as well as, if not rather than, the former. But that does not impose total uniformity in his account of the five senses, since he continues to recognise that some perception is of more distant, others of closer, perceptibles.

Again and more importantly, there is no question of his simply discounting the very great variety that exists, between different kinds of animals, in their sense apparatus and the ways they use it. True, many of the problems we might like an answer to do not get resolved, and some are not even considered. Reflecting, in part, his own views of where the important issues lie (for example, in the contrasts between humans and other animals), he discusses some topics in a cursory, even a minimalist, fashion – as for example the nature of the connections between the senses and the heart, where, once he can propose some link, the matter is not pursued further (though to be sure it would have been difficult to do so in the continuing absence of any conception corresponding to that of the sensory nerves).[6] But while that betokens a certain impatience with detailed analysis on some problems, on others that is cer-

---

[6] For the detailed arguments on this point, I may refer to my 1991, ch. 10, pp. 230ff.

tainly not the case. There is no question of his not appreciating that detailed investigation is necessary – no question, either, of his not then carrying it out – on such matters as the different varieties of the 'fleshy' and the 'bony' in the different kinds of bloodless animals, differences both as between the different main groups and indeed within them.

Nor, as the case of the sea-urchin shows, is there any question of his refusing to grant exceptions when he encounters what appear to him to be such, even when such exceptions pose a substantial threat to important general theories.

That he did not succeed in resolving all the problems of the varieties of perception – to his own satisfaction, let alone to what might be ours – is in no way surprising. That he has so much to tell us about its varieties gives us a very fine example of a distinctive feature of his style of exploration, namely the sense of the interdependence of philosophical analysis and detailed empirical investigation.

CHAPTER 7

The unity of analogy

Aristotle's concept of analogy plays several distinctive and important roles in his metaphysics and his zoology especially. I am not here referring to his use of particular analogies and comparisons, several dozens of which figure prominently throughout his work (Part II of my 1966, pp. 362–80, investigates a selection of these). I mean 'analogy' in the stricter sense Aristotle has in mind when he speaks of *to analogon* or *analogia*, where the root meaning is proportion, as in the relationship as A is to B, so C is to D. In such four-term proportional analogies, what is claimed is that the relationship within each pair is the same, a sameness distinct from sameness in number, sameness in species and sameness in genus, and labelled, precisely, sameness by analogy.

Aristotle uses that concept in two principal contexts, and at least one subsidiary one, particularly. I shall call these, purely for the sake of convenience, from their primary sphere of application, the 'metaphysical' use, the 'zoological' one and the 'metaphorical' one.

To deal first with what will be, for our present purposes, only a subsidiary question, the 'metaphorical' use as in analogical metaphor. *Metaphora*, transfer, is extensively criticised by Aristotle (as we shall see in chapter 10 below) and is condemned in definition and in reasoning in general. However, when he allows it a positive role in discussing style, it is particularly analogical metaphor that he praises (*Rhetoric* 1410b36ff., cf. 1407a13ff., 1412b34ff.). In one of the not terribly felicitous examples he uses on several occasions, the cup is to Dionysus as the shield is to Ares: so the shield is the cup of Ares. Or he attributes to Pericles the remark that when the city lost its young men, it was like the year losing its spring.

The relevance to us of this first use is that it shows that Aristotle is very much aware that some proportions express merely a 'metaphor', that is the transfer of a term from its appropriate domain to another one in which it is no longer used strictly, *kuriōs*. Given his strictures on the use of *metaphora*, the question we have to ask is whether, in his own positive and constructive deployment of analogy, in metaphysics or in zoology, for instance, he has justification for distinguishing this from the kind of metaphor or transfer he condemns.

The two principal uses of analogy he makes are, as I called them, the metaphysical and the zoological. The first is seen in the explication of certain high-level metaphysical concepts, such as potentiality and actuality, form and matter, or more generally the causes, principles or elements of things.

Faced with the difficulty of elucidating such concepts, Aristotle has two main resources. The first is what has been much discussed under the rubric of 'focal meaning'.[1] The canonical illustration is the term 'healthy', said primarily in relation to health, but also derivatively of signs of health (as when we speak of a 'healthy' complexion) or of what promotes or preserves or restores health (as when regular exercise, or a kind of climate, or a medicine are said to be healthy).[2] 'Healthy' is not to be understood and explicated in the same way when said of a climate or of exercise as when said of a patient who has recovered from illness (the primary use). But the term is not merely ambiguous or homonymous. *All* the other uses are connected with a primary one in relation to however we define health itself. The term is not said *kath' hen* but *pros hen*, that is it is not said 'according to' a single item, but rather 'directed at' a single item.

Aristotle's second resource is proportional analogy. Though these two are often associated, proportional analogy is not a *kind* of focal meaning, so much as an alternative to it. At least that is the way it is presented when he considers – very briefly – the analysis of the term 'good' in *Nicomachean Ethics* 1096b27ff.

How proportional analogy may be used emerges, for example, from such texts as *Metaphysics* Λ 4. There he says that the causes and principles of different things are different, but, in a way, 'if one speaks universally and by analogy', the same for all (*Metaph*. 1070a31ff.). Thus although every object or event has its own form,

---

[1] Owen presented the classic analysis in a series of papers dating from 1957, 1960 and 1965, reprinted in his 1986. Among the most notable contributions to the on-going debate have been Barnes 1971, Hamlyn 1977–8, Irwin 1980–1, Fine 1982.

[2] See *Metaph.* 1000a33ff., cf. *Top.* 106b33ff., *Metaph.* 1030a35ff., 1060b37ff.

privation and matter, and in that sense the causes are different, they are in another sense the same – that is by analogy – in that they are, precisely, the formal cause, its privation, and the material cause. At 1070b17ff. he illustrates this with colour and with day and night. In colour, there is white, black, surface, and in day and night, light, darkness, air. By the end of the chapter, he has added the efficient cause and given two more examples of how the causes are analogically the same: in medicine, there is health, disease, the body, and the medical art as efficient cause, and in building, the relevant form (it might be a house), the corresponding privation (the disorder in the materials before the house is made), the material itself (e.g. bricks) and the house-building art.

Again in *Metaphysics* Θ 6, 1048a35ff., potentiality and actuality are similarly explained. We should grasp the analogy, he says: just as that which is actually building is to that which is capable of building, so that which is awake is to that which is asleep, and what is seeing to what has its eyes closed but has the faculty of sight, and that which is differentiated from the matter to the matter, and what has been worked up to that which is not yet worked up. In this instance he recommends grasping the analogy *in distinction to* attempting a definition.

The second main use which we shall be considering in detail later in the course of this chapter is that which identifies a level of difference. The accounts of both 'one' and 'different' in *Metaphysics* Δ introduce the main distinctions. Thus 1016b31ff. says: 'some things are one numerically, some in species, some in genus, some by analogy. Those things are numerically one of which the matter is one: those things are specifically one of which the definition is one: those things are generically one which belong to the same category; those things are analogically one that have the same relationship as two other things have to one another.'

Most often such levels of difference are illustrated by, and applied to, the differences in the parts of animals, or their activities (which is why I have labelled this the 'zoological' use). Some differences between parts are merely a matter of 'the more and the less', as happens between different species in the same genus.[3]

---

[3] This is not to imply a commitment to there being determinate levels at which 'species' and 'genus' are used. On the contrary, as is by now generally agreed, those terms are used at many and various levels, though they are, occasionally, used to express a stratified relationship, as in the metaphysical account of grades of unity, 1016b31ff. quoted above.

Some birds have long beaks, others short, for instance. But some differences are such that the parts are the same only in analogy. *HA* 486b19ff. gives bone and fish-spine, nail and hoof, hand and claw, feathers and scales, as examples.

All of the material I have so far referred to is very familiar indeed. Yet some of the problems it presents do not receive the attention they deserve. We shall be occupied mainly with three.

(1) How far do either the metaphysical or the zoological uses square with the requirements of the *Posterior Analytics*? Although the *Posterior Analytics* itself refers, as we shall see, to analogy, especially in II 14 and 17, it is not clear how its own use is thought to tally with the recommendations it sets out for definition, for demonstration, for understanding (*epistēmē*, 'science') itself. How far does the *Posterior Analytics* set out an adequate theory to encompass analogy, and how far do the metaphysical and zoological uses conform to what the *Posterior Analytics* leads us to expect?

(2) What are the constraints on the use of similarity by analogy in practice in zoology? How would Aristotle rebut the charge that what he identifies as analogues is merely what his overarching theories *dictate* should in some sense be the same? He repeatedly invokes what is analogous to blood in the bloodless kinds, what is analogous to flesh, to the heart, and so on. But what precise justification is there for this?

(3) Having explored some of the difficulties in the metaphysical and zoological uses, we may ask what unifies the concept of 'analogy' for Aristotle and return to the distinction between his constructive use and analogical metaphor. Is analogy too, as we suggested for demonstration, a case of a pluralism of theories and practices, adapted, more or less coherently, more or less systematically, to different contexts?

(1) The first set of problems can be readily identified. Each branch of understanding has, Aristotle insists (for example *Posterior Analytics* 75a42ff.), its own genus, and the demonstrations it aims at have their own proper principles (for example, 71b22f.). Those principles include, especially, definitions, the standard form of which, for substances at least, is by way of a genus and its differentiae (for example, 96b36ff.). For valid deduction of any kind, the terms must be used univocally: moreover the premises that yield demon-

strative syllogisms have to meet the rigorous requirements set out in *Posterior Analytics* I 2, 71b20ff. (cf. above chapter 1 at p. 12) namely they must be primary, immediate, better known than, prior to, and explanatory of, the conclusions. In such premises, the terms must be predicated *per se* and universally, both *kata pantos* and *katholou* in the special sense he elaborates at 73b26ff., namely not just true of all the subject, but belonging to it *per se* and *qua* itself, the so-called commensurate universal.[4] Having internal angles equal to two right angles is not in this sense a commensurate universal with regard to isosceles triangle: though true of all isosceles triangles, it is not true of them *qua* isosceles. The commensurate universal in this case will be triangle.

But how, if sameness and oneness by analogy *transcend* the genus, as in both the metaphysical and the zoological uses, does that not breach those requirements? How can there be understanding, in the strict sense there defined, of any such cases, if understanding is limited to the genus? How can there be demonstrations, if these proceed from principles that include some that are not proper to the branch of understanding in question? How can there be deduction, even, if the analogical use of terms departs from the strictest *kath' hen* model, as is implied for focal meaning at least in *Metaphysics* 1003b12ff cf. 1048a35ff.? How can there be definition, on which understanding depends and by means of which it is attained, if definition requires a proper genus to be differentiated?

Some alleviation, of course, is possible. It is by now well known that the term *genos* can be used quite loosely and may refer to kinds at very different levels of generality. When the *APo.* insists that each branch of understanding has its own determinate *genos*, that should certainly not conflict with the study of certain properties or parts that are only analogically the same across different species of animals. The feathers of birds and the scales of fish are not the same in *genos* in one sense, but they both fall perfectly satisfactorily within the purview of an inquiry into animals, where animal provides the determinate *genos* to meet the *APo.* condition.

As for the metaphysical use too, Aristotle would not, in any case, claim that potentiality and actuality as such (for instance) consti-

---

[4] The question as to whether or how far Aristotle is committed to a notion of such a 'commensurate universal' is, however, disputed: with Barnes 1975, pp. 247f., compare Inwood 1979.

tuted the proper subject-matter for a *specific* branch of understanding or the topic of the demonstrations aimed at in such. Like the distinct but related difficulties to do with the high level use of focal meaning in the study of being *qua* being, quite what kind of study, exactly, the analogical investigation of the principles and causes is, poses a problem. But whatever answer we give, such topics were never to be treated as precisely the same type of 'branch of understanding' as those targeted in the *Posterior Analytics*, such as geometry, arithmetic, harmonics and so on. So while the analysis of potentiality and actuality, for example, fails the *APo.* test, of having a determinate subject genus, there was, in such cases, never any question that they should have to pass it in the first place.

Then as to the requirement that the principles should be proper, Aristotle allows, first, that it is possible for subordinate branches of understanding to share the starting-points of others they are subordinate to (*APo.* 76a11–15) and he goes on to point out, 76a37ff., that while some of the principles used in demonstrative understanding are special, others are common, *koina*, though common, precisely, 'by analogy'.

He illustrates this not, as one might expect, with the common axioms such as the Law of Non-Contradiction or the Law of Excluded Middle, that have to be understood if *any* knowledge is to be achieved, 77a10ff., a26ff. At 76a37ff. he offers as an example not a universal axiom, but one that spans several fields nevertheless, namely the equality axiom, take equals from equals and equals remain, common to both geometry and arithmetic. But while that provides *one* good example of a principle 'common by analogy', just how many such there will be, whether among the axioms, the hypotheses or the definitions, is not clear.

With definitions, indeed, we come to the main difficulty. Although definitions as such are indemonstrable, the middle terms of proper demonstrations, setting out the causes, correspond to the definitions of the essences in question. In some of his chief discussions of this issue, in *APo.* II, Aristotle sometimes takes events, rather than substances, as his examples, but they illustrate the two chief points we are concerned with clearly enough, namely (1) the need for a genus, and (2) the fact that the middle term sets out the cause.

Thus thunder is defined as the extinction of fire in the clouds.

The demonstration of the cause of thunder will proceed by using the knowledge that the definition contains: the noise occurs because the fire is extinguished in the clouds (*APo.* 93b7ff.). Similarly, the definition of an eclipse of the moon is the deprivation of light through the obstruction of the earth, while the corresponding demonstration shows that the eclipse occurs because the earth obstructs the light (90a15ff.).

So on the *Posterior Analytics* model, demonstration and definition are crucially interdependent. But where there is no proper genus, there is no proper definition. Moreover it is not just with metaphysical analogies that genus boundaries are crossed – and in this case the ones that are relevant to proper definition – but also with zoological ones too. Indeed as we saw, at *Metaphysics* 1048a35ff. the analogical explication of potentiality and actuality is offered as an *alternative* to definition.

Here too palliatives may be possible, but how far they succeed in bringing zoology satisfactorily within the compass of the ideal set out in the opening chapters of *APo.* 1 is doubtful. Some attempts suggest an application of the requirements to zoology only at the price of modifying or relaxing the requirements, though whether when that happens we are left with an adequate theoretical basis for the use of analogy remains problematic.

It is true that some broadening of the scope of application of syllogistic in general is apparent already in the *Prior Analytics* when Aristotle introduces an analysis of syllogisms with propositions that hold 'for the most part'. This looks ideal from the point of view of the study of nature, which deals with what is true 'always or for the most part'. But there are, as I have noted (chapter 1 at p. 13 and additional note to p. 22) very considerable problems. Briefly, if we take 'for the most part' in what might seem the most obvious, statistical, sense (where 'most' is a matter of the majority of cases, and contrasted with 'in *all* cases') that falls foul of the form of the claim made in *APo.* 1 27, 43b33ff., that when problems are 'for the most part', the syllogisms will consist of propositions that are – 'either *all* or some of them' – for the most part. But if 'for the most part' is taken as 'most', then the problem is that if most Bs are As, and most Cs are Bs, it will not follow that most Cs are As: it will not follow indeed that any are.[5]

---

[5] Various alternatives to the 'statistical' reading of 'for the most part' have been canvassed, for example that it should be treated as a temporal operator ('not always') or a quasi-

Unity of analogy

The ambition, in the *Posterior Analytics*, to extend not just syllogistic reasoning in general, but demonstrations in particular, to such fields as zoology and botany is obvious not least from the examples from such fields that punctuate the account, especially in book II. Two chapters in that book (14 and 17) in particular directly raise questions to do with terms used analogically, and in the first of these we are given one of the standard zoological examples, namely that of bone, fish-spine and pounce (the bone-like structure in certain cuttlefish). This too looks most promising, but the question is, does Aristotle succeed *either* in bringing such analogies within the orbit of the general theory of scientific understanding set out in *APo.* I, *or*, at least, in providing a satisfactory alternative framework for their use?

II 14 deals with the selection of problems, that is the connections to be proved, and proceeds first, at 98a1ff., with cases where there is a common genus, namely animals, where we are to consider first what properties belong to *every* animal, then what to the main groups in turn, for example birds. This is the standard procedure with which we are familiar. Then at 98a13ff. the difficulty, that a common characteristic may exist but have no name, is mentioned, but that too causes no real obstacle. With any such characteristic, named or not, we should consider both what this characteristic is consequent upon (*akolouthei*) and what follows from it (*hepetai*) in turn. In the example given, having a third stomach (*echinos*) and teeth in one jaw only (i.e. incisors only in the lower jaw) 'follows from', or belongs to, being horned. It emerges from the discussion in the *de Partibus Animalium*, consistently with this, that having horns implies having no incisors in the top jaw (663b35ff.), but the converse does not hold (the camel has no horns, but no incisors in the top jaw, 674a31ff.).

Then at 98a20ff. he says that there is another method of selecting, namely by analogy. It is impossible to grasp one and the same thing by which to call pounce, fish-spine and bone: yet there are properties that follow from these 'as if there is a single nature that is suchlike'.

The task of investigating the relationships (here called 'consequent upon' and 'following from') between groups of characteristics here extends to characteristics that are the same only 'by

---

modal operator ('not necessarily') or that it corresponds to some admittedly unanalysed notion of what holds 'by nature'. Yet difficulties remain: cf. Mignucci 1981.

analogy'. The lack of a common name, as in the immediately preceding text, 98a13ff., should be no deterrent. Yet in this case, where there is sameness 'by analogy', Aristotle adds '*as if* there is a single nature'. How is this 'as if' to be taken?

This suggests a difficulty that can initially be presented in the form of a dilemma. If there is *indeed* a common nature (nameless though it be) then there will be no problem. The case falls into line with those where there is a common genus. The analogical unity is *not just* a matter of a relationship (as A is to B, so C is to D, and E is to F) since, *in addition* to the relationship between the pairs being the same, A C and E all share a common nature: they fall under a higher *genos*.

But if not, if there is no such common nature, and the pairs have nothing in common but the fact that they stand in a similar relationship to one another, then where would be the justification for thinking of properties following from these, or these following others?

Of course 'unity by analogy', it might be urged, is designed precisely to capture the type of case where there is no unity of *genos*, and so no common nature of that kind, and yet a unity nevertheless, for example certain common capacities or a common function. Yet that does not resolve the problem of whether the analogical relationship *by itself* is sufficient to justify any assumptions concerning the connections between the relata.

The difficulty I have in mind can be brought out by considering some other cases of pairs of things that are *just* the same 'by analogy'. Take one of the metaphysical uses we mentioned before, the elucidation of potentiality and actuality in *Metaphysics* Θ. How would one set about 'selecting the problems' in the case of 'building' and 'seeing'? These two actualities both stand in a certain relationship with a corresponding potentiality, what is capable of building, what is capable of seeing. There is no doubting their credentials as a case of sameness 'by analogy'. Yet since building and seeing share no common properties, they will not yield any of the connections to be proved within a branch of understanding such as the *Posterior Analytics* desiderates. In the other metaphysical example, from *Metaphysics* Λ, where Aristotle speaks of the principles and causes of things being the same by analogy, the same thing applies. The form, privation, matter, efficient cause relationship is the same in each of the cases, but there are no common properties that link, for instance, the various material causes – bricks and the body – in the examples from building and medicine that he gives.

A fortiori, one might add, analogical metaphors – of the types illustrated above, the cup of Dionysus and the shield of Ares, and the loss of youth and the loss of the spring of the year – are cases of analogies in the strict sense of proportions, but again could not serve as guides to any scientific connections to be proved.

These examples show how problematic it would be *just* to rely on analogical sameness for problem selection. There are, undoubtedly, scientific connections to be investigated between things that are one by analogy, but their exhibiting that unity is no guarantee as to the connections.

The same point applies with even greater force to the second chapter that mentions, even more briefly, the exploration of what is the same analogically, namely II 17. According to 99a15f., in the discussion of causes, things that are the same by analogy will have middle terms that are also analogical, here opposed on the one hand to merely equivocal terms, and on the other to terms that express a generic connection. The status of these analogical middle terms between these two options is not further clarified. But if we are talking about middle terms that pick out the essences, essences will only be found if the analogical relata have more than just the relationship in common.[6]

Within the *Posterior Analytics* itself, therefore, there is a problem of consistency to be faced. The normal requirements on definition, essence, demonstration (in the strict sense there specified) all depend on the notion of the genus, and cannot be met by what is the same *not* in genus, *but only* by analogy. Yet the ambition to extend the analysis to cover what is the same by analogy is there, even though there is no clear sign, in the passages in question, as to quite how this is to be achieved *if* the standard requirements remain in place.

Did Aristotle think that his references to analogy could be incorporated straightforwardly into his general account of definition and demonstration in the *Posterior Analytics*? Some have maintained that that must be the case. Yet from the arguments I have given, it

---

[6] This point remains valid, even if the middle terms are, as Lennox has suggested, treated as disjuncts.

seems that for that to be accomplished one or other of two things has to happen. *Either* quite drastic modifications have to be made to those strict, normal, requirements – to allow for definition other than through the differentiation of a given genus, and to permit demonstration through middle terms other than those that pick out essences that are specifications of such a genus. *Or* the notion of the analogical relationships in question has to be clarified, restricted indeed to those cases where there are essential properties shared by the relata. In the latter case, the theoretical basis for the use of analogies still needs very considerable further elaboration – a point suggested, in any event, by the desperate brevity of his references to them which seem, in context, little more than throwaway suggestions.

The alternative, much weaker, line of interpretation would have it that these references to sameness by analogy are intended as no more than recommendations to *explore* such cases, not as an alternative to the model of definition and demonstration that has been used in the body of the treatise, but as a preliminary to the application of that model. Thus the discussion of problem selection in II 14 would be seen as only the *first step* towards identifying the connections to be proved. Analogical relationships are worth examining in that context, since they may reveal that there *are* common attributes shared by the relata. Once it is established that pounce, fish-spine, bone all have a common function, then the way is clear to attempt to investigate what that function is consequent upon and what follows from it. But the analogical relationship by itself is no guarantee that such a common function will be found.

It has to be said that that second line of interpretation, too, goes appreciably further than the statements in the text, as would a similar argument applied to the admittedly highly elliptical passage in II 17, 99a15f.

Yet whatever we make of those problematic chapters, the conclusion of this first part of our study must be that the metaphysical and zoological uses – as usually presented – sit uneasily with the model of *epistēmē* that the *Posterior Analytics* is mostly concerned with.

(2) Our next task is to review the actual practice in the zoology, where the use of sameness by analogy is a good deal more complex and varied than we have so far been able to indicate. Thus

the range of types of case to which sameness by analogy applies include: (1) the parts of animals, both (1a) homogeneous ones, as in the chief zoological examples we have considered so far, such as pounce, fish-spine, bone, and (1b) heterogeneous ones, as in hand and claw (*HA* 486b20) or the chest and what is analogous to it (*HA* 497b32ff.); (2) vital functions, including (2a) those relating to nutrition, such as digestion (cf. chapter 4) and certain features of sexual reproduction, as when menses in females are said to be analogous to semen in males (*GA* 727a3f.), and (2b) those relating to cognitive faculties, such as the varieties of perception, notably touch (cf. chapter 6); (3) the characters or activities of animals, as when animals are said to manifest what is analogous to art, *techne*, or wisdom, *sophia*, *HA* 588a28ff.; (4) and a variety of special cases, as when the generative substance in the *pneuma*, in semen, is said to be analogous to the fifth element, the element of the stars, *GA* 736b35ff.; or when the generation of bees is said to incorporate an analogy (a matter of a series of terms standing in a proportionate progression to one another, *GA* 760a12); or when the relationship between plants and earth is analogous to that between testacea and the sea, *GA* 761a26ff.

It is impossible here to do justice to the richness of this material: nor can we tackle all the individual problems of interpretation it presents. But we may concentrate discussion on four topics in ascending order of importance, first the question of namelessness – the defaults of the natural language – second the fluctuations in the analysis of particular cases, third the adequacy of the criterion offered when what is analogous is explained as what has the same function, *dunamis*, and fourth the extent to which the analogies that Aristotle deploys are not similarities he discovers but postulates answering to the needs of his general theories.

The lack of a name for what items that are analogous have in common was already mentioned in *APo.* 98a21f. But namelessness, in the zoology, takes two forms. First there is the type of case represented by pounce, fish-spine and bone, where the problem is the lack of a term for what these have in common. But secondly and, as we shall see, more importantly, there are many instances where what is lacking is a term for *one of the* analogues. One example has just been cited: chest or what is analogous to it (*HA* 497b32ff.). But there are many others, most notably blood and what is analogous to it, flesh and what is analogous to it, heart and what is analogous to it. True, in some instances there is a name for the part that serves as the analogue, though that may or may not reveal its function. Thus in the cephalopods there is the organ called the *mutis* that Aristotle identifies as the analogue of the heart, *PA* 681b26ff. (cf. below). But more often the part is nameless – especially in the lower kinds of animals – and Aristotle refers to it simply as what is analogous, in those creatures, to the given named part familiar from the higher animals.

We may note the evidence this yields for Aristotle's recognition of the shortcomings of the natural language he uses, a feature that can be exemplified elsewhere in his zoological researches, though, unlike on some other occasions, those shortcomings do not here lead him to produce many new coinages. The point he is interested in, in any event, is not their names, but the relationship. Calling a fluid the analogue for blood, for instance, may be cumbersome, but it is immediately informative about the presumed role of the fluid in question.

The first of the main difficulties we encounter, in Aristotle's zoological use, relates to the variations in the analysis he offers of some phenomena or processes. What we are told about bone and cartilage in *PA* II 8f. is only a mild case. At *PA* 655a32ff. (cf. also *HA* 516b31f.) the nature of bone and cartilage (*chondros*) is said to be the same: they differ by the more and less. Yet when he considered the analogues of bone in fish at *PA* 653b35ff. he said that in some fish there is fish-spine, in others cartilage. So within the same creature cartilage and bone may differ from one another only by excess and defect, but across kinds may, rather, be analogous.

Then at *HA* 516b3ff. the usual distinction between 'by analogy' and by 'more and less' seems to be eroded, if our text is correct. The bones of footed vivipara are said not to differ much, but 'by analogy, in hardness and softness and in size' – where the last three differentiae would normally be matters of differing by 'more and less'. By the end of the chapter, 516b12ff., comparing what other blooded animals have, we have the usual view that fish have only what is analogous to bone, though the complication there is that the selachians are cartilaginous-spined, the oviparous fish have regular fish-spine.[7]

---

[7] Peck, 1965, p. lxiii, sees this as a further 'analogy' *within* the fish, viz. as between the cartilaginous and the oviparous ones: but it may be that the point is the more usual one that fish do not have bone but what is analogous to it.

Unity of analogy

Again what is said about male and female in plants and testacea poses a problem. At *GA* 715b19ff. he says that male and female are spoken of, in those kinds, only 'according to a similarity and by analogy, for they exhibit a small difference of that sort.' Yet at *GA* 731a1ff., 741a3f., he asserts that there is no separation of male and female in plants, and similarly at 731b10f., *HA* 537b24f., for the testacea. Again at *HA* 537b22ff. the terms *tiktein* (bear) and *kuein* (conceive) can only be said 'according to a certain similarity' of animals where there is no male and female: but at 538a18f. he allows *tikton* (capable of bearing) and *gennōn* (engendering) of testacea and plants, but not *ocheuein* (copulation).

Apart from questions to do with the strict consistency of Aristotle's account, the issue that such usages raise is when we are dealing just with what Aristotle would call *metaphora*, transfer, that is when a term that has a proper usage in one domain is transferred to another (cf. chapter 10). There can, as we noted, be *good* metaphors, that is ones that are effective and vivid, indeed this is especially so when they express a proportion (*analogia*). While he is often critical of *metaphora* in his natural science, he allows that some can be well found. Thus at *GA* 784b19ff. he endorses the comic metaphor, where white hair is said to be the 'mould' or 'hoar-frost' of old age, but he gives an elaborate rationale for this, 784b8ff. Greyness in hair is due to putrefaction or a failure to concoct, but is like mould since mould too is the putrefaction of earthy vapour and mould too is white. Moreover mould is the converse of hoar-frost, since mould is putrefied vapour while hoar-frost is congealed vapour.

Here then there are certain underlying similarities that legitimate the 'transfer'. But in another case, that of the horns of certain non-viviparous animals, discussed at *PA* 662b24ff., the situation is very different. Only the vivipara have horns, he says there, but 'horn' is used 'by similarity and metaphor' with regard to some other animals, though the parts in question in no case perform the task (*ergon*) of horns.

Both 'by similarity' and 'by analogy' can be used *both* of cases of (mere) *metaphora* and of cases where there is an underlying common nature. That means that further analysis is needed to determine which type of case is in question, but the potential for slippage is obvious and it is not always the case that Aristotle removes all doubt as to his view of the relationship.

This takes us already to our third main question. One of the canonical texts that explains the contrast between differing by the more and the less and sameness/difference by analogy, namely *PA* 645b6ff., puts it first that some animals have a lung, but others something 'instead of it' (*anti toutou*), and then that some have blood, others 'the analogue that has the same function/capacity (*dunamis*)', with which we may compare *PA* 648a19ff which speaks of what has the same nature (*phusis*) as blood. The problem is how adequate an account of the similarities in question Aristotle provides and to what extent (our final question) do his zoological uses appear to be driven by the needs of his overarching theories.

On the first of those two questions, everything will depend, of course, on which *capacities* are the ones that count. Aristotle has, to be sure, no equivalent to the technical distinction between homology and analogy, between parts that are morphologically and genetically similar, and those that merely serve a similar function. He does have some interesting remarks to make about the bat's wings – which are anomalous in having feet and being 'skin-winged', *PA* 697b7ff., *HA* 487b22f. – but he treats them as wings, not merely their counterparts. Again hands and claws are 'analogous', for they generally serve the same function, e.g. *HA* 486b20, but in the lobsters the claws are not used for prehension, but rather in locomotion, *PA* 684a33ff. But as that example also shows, the adaptation of parts does not interest Aristotle as a sign of the different modes of analogy of function there may be, and their different origins: he is interested in these phenomena, rather, as indications of the anomalous, even deformed, nature of the animals in question.

In Aristotelian usage *dunamis* can refer to a potentiality of any kind, any capacity to change or be acted upon, and in some of the proportional analogies he proposes, no more than a passive capacity seems to be involved. Thus in the *Meteorologica* 387a32ff. when he is talking about the evaporation or fumes of different types he says that those of 'woody' substances are smoke and glosses this: 'I mean bone and hair and anything that is suchlike ... even though there is no common name, yet by analogy they are all alike.'

But far more often the capacities in question relate to more than just the characteristics of the material substrate. How much more is an interesting question with regard to the analogy proposed between menses and semen, e.g. *GA* 727a3. They are certainly the same 'by analogy' in one sense, namely that they are each the end-result of the process of concocting the useful nourishment, although females are defined by their incapacity to carry this through – viz. in the way males do, to produce semen (cf. chapter 4). Yet from the point of view not of how they are produced, but of what they can do, their own capacities, semen and menses play very different roles in generation, semen supplying the formal and efficient causes, menses the material. While Aristotle is prepared to talk of female seed, or the female spermatic residue, he qualifies those expressions with remarks that it is 'not pure' or 'in need of elaboration' (for example *GA* 728a26f., 737a28f., 766b12–16).

Although there is, potentially, very considerable indeterminacy in the capacities or potentialities that have to be the same to justify the proposal of a proportional analogy in zoology, in practice in many of the most important cases it is clear that the focus of Aristotle's interest is on the essential vital activities of the animals in question, and indeed of animals in general.

Thus three such cases are the analogues (1) to flesh, (2) to blood and (3) to the heart. The first of these has already occupied us in chapter 6 and to a lesser extent in chapter 3. Why it is crucial, for Aristotle's purposes, that an animal possesses flesh or some analogue is that – normally (cf. chapter 3) – animals are defined by the capacity to perceive, the basic mode of perception is touch, and the organ or the medium of touch is the flesh. He *might* have proceeded by a quite different route, that is if he had connected flesh with the faculty of locomotion: but he does not, and that is reflected in the very little he has to say about the musculature – whether in the *de Incessu Animalium* the *de Motu Animalium* or anywhere in his zoology.

On the other hand he recognises that flesh may be softer or harder, firm or spongy, 'dry' or 'wet', and he is concerned, as we have seen in chapter 6, with the difference between those animals with their flesh outside the bones and those with it, or its equivalent, inside. His solution, with regard to the insects, is to say that their flesh is neither shell-like, nor flesh-like, but intermediate (*HA* 532a31ff.) or intermediate between the fleshy and the bony (*HA* 523b15f.) or, in yet another formulation, more fleshy than bone and more bony and earthy than flesh (*PA* 654a27ff.). In line with his theory of the role of flesh in the primary cognitive activity that defines an animal as an animal, part of his research into flesh and the flesh-like is directed to confirming that some such part exists. Even so, as we noted in chapter 6 that general theory does not prevent him from recognising exceptions. Thus at *PA* 679b34 he states that the sea-urchin (*echinos*) has no fleshy part (yet cf. 680a5ff., glossed in turn at 680a11f.) and at *HA* 528a6f. it is cited to show that some testacea have no flesh.

The interest in blood and its analogue is even more basic, since it is directly connected with the nutrition of animals. The food any animal consumes has to be digested and the result of the first stage of that process is blood, from which, in the blooded animals, all the other parts of the body are produced (cf. chapter 4). So blood is essential, the fluid on which life itself depends (e.g. *HA* 489a20ff.). Moreover in the blooded groups, differences in the blood are used, as we have seen (chapters 2 and 4), to account for a wide variety of physical and psychological differences, varying from the difference between producing fat and producing suet, to variations in the characters and intelligence of the animals concerned.

But what then of the other kinds of creatures, Cephalopods, Crustacea, Testacea and Insects? Despite the fact that they are *called* the bloodless animals (*anaima*), they must nevertheless have an analogue to blood. That is stated in general terms repeatedly. Yet in practice, in the detailed accounts of the internal parts of these kinds, whether in *HA* (especially IV 1–7) or in *PA*, he tells us much less about the nature of that analogue to blood than one might have expected in view of its importance. This is particularly striking since, as Peck pointed out, it is only *red* blood that Aristotle recognises as blood. Here is a case where Aristotle, for once, underestimated the strength of the analogy he proposed, for first red blood exists in some of the animals he considered 'bloodless', and secondly others of Aristotle's 'bloodless' creatures have blue or green blood that serves a similar respiratory function.[8]

Thus we are certainly told that the bloodless kinds have no viscera, and again no blood-vessels, no bladder and so on, since they have no blood (*PA* 678a26ff., a36ff.). Their not having blood is also invoked to explain their being colder and more fearful (e.g. *PA* 679a25f.), while their having (at least) an analogue enters into

[8] See Peck 1937, note to *PA* 645b9.

Unity of analogy

the account of their producing seed (*GA* 726b3ff., 728a20f.). But while we are given careful discussions of such questions as whether the bloodless kinds have teeth, or tongues, what kind of stomach they have and the kind of residue they produce (e.g. *PA* 679b10ff., 680a20f., *HA* 529a11f.),[9] we have comparatively little on the nature of the analogue of the blood. In the account of the crabs there is a cryptic reference to pale juice in the trunk of their body, though this is not further explained, *HA* 527b28f. The pale fluid in the crayfish, at least, *HA* 527a17ff., comes in a description of their reproductive parts. The analogue to blood is said to be imperfect and to be *like* fibre (*is*) or serum (*ichōr*) at *HA* 489a21ff, cf. 511b4: yet fibre and serum are themselves forms or products of blood in the blooded animals, *HA* 515b27ff., 521b2f., *PA* 651a17ff. (cf. chapter 4 p. 87).

One text that does refer more clearly to the particular character of the fluid in one of the bloodless kinds is that which identifies the *mutis* of cephalopods with the heart (*PA* 681b26ff., cf. 524b14ff.). It does so first on the basis of the position of the organ in the centre of the animal (where Aristotle expects the heart or control centre of the animal to be), and secondly because of the sweetness of the fluid it contains, a sign that it has undergone concoction and is 'bloodlike'. From Aristotle's own point of view, that second consideration is a tricky argument. He accepts that in the blooded animals blood is present throughout the body and not just, even though admittedly especially, in the heart. It exists for instance in the liver – which is indeed the role that modern zoology would assign to what Aristotle identifies as the *mutis* in cephalopods.

The way Aristotle proceeds in his accounts of the analogues of the heart is most revealing of all. Again the driving force is a battery of arguments about the necessary functions fulfilled by the heart – and so equally necessarily by its analogue in the bloodless kinds. Several texts identify the three main essential parts of animals as that by which food is taken in, that by which it is discharged, and what is in between, where the controlling principle is located, *De Juventute* 468a13ff., *PA* 655b29ff. His doctrine of perception is that this is basically a function of a common *aisthētērion*. So the heart is the seat not just of perception and imagination, but

---

[9] In these texts, the *mēkōn* of testacea is sometimes treated as equivalent to a stomach, sometimes as itself a residue.

of life itself. He has, of course, empirical evidence for the fact – as he believes it to be – that the heart is developed first in the growing embryo, especially his famous investigation of the growth of the embryo chick *HA* vi 3, 561a4ff., and he would also no doubt claim that the connections between the sense-organs and the heart are well established.[10]

It was Aristotle's view that an animal, *as* an animal, had to have a centre of its vital functions. But if that meant that there *had* to be some analogue to the heart in the bloodless kinds, *what* part served that role was, in many cases, far from obvious. Aristotle evidently *infers*, rather than observes, the analogue to the heart, as he does in the case of the *mutis* of cephalopods. But guided by certain general views about the value and importance of the centre, as well as by the analogy of the position of the heart in the blooded groups, he assumes that the analogue will usually be found there.

This is indeed stated as a general rule at *PA* 681b33ff. In the testacea, he there admits, the part in control of perception is hard to make out. But it should be looked for between the part where the food is taken in and the part where the residue is excreted (that is, in his view between what is defined as the upper and the lower parts of the animal) and in those kinds that move about between right and left (since right and left relate to the principle of locomotion in animals). So too in the insects, he continues, 682a1ff., the controlling part is located between the head and the cavity where the stomach is.

Here then we seem to have a prime example where *a priori* principles dictate not just the programme of research, but its results. However, as in the case of the analogue to flesh, considered above, and as, more strikingly, in the matter of perception as the defining characteristic of animals, according to the results of our study in chapter 3, so here Aristotle's general principles do not stop him from recognising exceptions. Normally the control centre is single, and in the centre of the animal and responsible for all its main vital functions. Yet some animals continue to live even when they are cut up.

The point is repeatedly made. *HA* 531b30ff. states that *all* insects live when cut up except those that are excessively cold or very small and so quickly chilled. With the middle part, the head and

---

[10] Cf. above, chapter 6 at p. 128.

the stomach can live, but the head not without it. Long, many-footed insects live for a long time after being cut up, and the portion cut off can move in either direction – towards the cut or the tail, as happens in the so-called scolopendra.

IA 707a24ff. starts with the denial that any of the blooded animals can live for any length of time when cut up, but it too recognises the same facts about insects and especially the many-footed ones as *HA* and offers an explanation (707b2ff.): they are like a continuous body made up of many animals.

Finally *PA* 682b20ff. offers a further rationale of creatures having 'insected' bodies, that this allows them to curl up and escape injury. But there is not just this final cause at work. It necessarily belongs to their being to have many principles, *archai*, in which they resemble plants. Noting that plants too can live when cut up, he remarks this difference that insects only survive for a time, and in plants you can get two or more perfect specimens from a single divided individual.

Of course the insects in question are very lowly creatures. Nature aims for a *single* principle, and admits a plurality only, as it were by default (*PA* 682a6ff., cf. 667b21ff.). But it is remarkable that once again Aristotle's application of the principle of analogy is saved from dogmatism thanks to a further general theory of his, namely the hierarchy of the animal kingdom.

(3) We have explored just a few of the manifold applications of the notion of sameness by analogy in the metaphysics and the zoology. But how, we may ask in conclusion, are these various uses to be related and is any unity to be detected in them? Granted that analogy is not a substantive doctrine, can we see it even as offering a unified methodological procedure?

First, due weight must indeed be given to that variety. In metaphysics, proportional analogy provides one of the main ways in which Aristotle elucidates extremely abstract and fundamental ideas. Within zoology, analogical relationships take many different forms. Some enable him to develop and apply, across the entire animal kingdom (or nearly so), certain central doctrines about the vital functions that make animals the animals they are, both severally and in general. Others follow up incidental comparisons between phenomena or processes some of which may even be suggested by the metaphors used by the comic poets.

What unifies a very large number of the analogies we have surveyed may be said to be their *heuristic* value. True, how this is cashed out varies in different cases. When Aristotle is already confident of his overarching theories, the notion that a part or organ in certain groups of animals has an analogue in others amounts to a recommendation to find the analogue. On other occasions, the perception of an analogical relationship itself stimulates exploration of whether there is a common nature, and if so, if an explanandum has been identified, to finding the explanation.

Crucially, the actual analysis of the items compared and the relationships between them varies. The initial appreciation of a proportional analogy, or the suggestion of the possibility of one, do not dictate the final verdict on the phenomena thus brought together, and this in two respects. First, the status of the similarity apprehended or revealed varies, and secondly, the overall thesis they may support (when they support one) may be subject to exceptions. Even his most cherished theories about the vital functions are not exceptionless, and by that I do not mean just that there may be individual exceptions, but whole kinds or groups of creatures that do not exactly fulfil the conditions laid down by the theories.

The very fact that proportions may be no more than 'metaphorical' warns us that the mere discovery of a similarity in a relationship does not tell us anything fundamental about the members of each pair of relata. Any such discovery will need further scrutiny, which is, indeed, often precisely what we find Aristotle undertaking, and the outcome of which – again more often than many commentators would expect – is far from a foregone conclusion. In some cases, indeed, we have reason to doubt that Aristotle has reached a *definitive* conclusion at all.

What this topic reveals is another aspect of the openendedness that characterises important areas of Aristotle's work. The possible applications of his metaphysical notions, of potentiality and actuality, of form and matter, are capable of further development. There is no question of Aristotle giving a final list of those applications. Indeed we see him extending those notions, as he uses them, both in metaphysics and elsewhere, and including into problematic areas that, rightly, puzzle us. Such is the case when, for example, the matter side of the form-matter dichotomy is illustrated not just with the physical matter of a table, but also the intelligible matter of mathematical entities and the genus in relation to its differentiae. Similarly, his investigations of animals exhibit a fruitful tension, between the need to see what *connects* them as all members of the multifarious animal kingdom, and the recognition of their very *variety*, a variety that he explores, so often, by way of the proposal and examination of proportional analogies.

CHAPTER 8

Heavenly aberrations:
Aristotle the amateur astronomer

Aristotle's astronomical account, in *Metaphysics* Λ 8, contains a number of famous cruces that have so occupied the attention of commentators that they tend to miss some important strategic questions that that account might be thought to raise. The two issues that receive most discussion are: (1) has Aristotle miscalculated the number of spheres, direct and retroactive, he needs?[1] When he considers the possibility, at the end, of not accepting Callippus' proposal of certain extra spheres for the moon and sun, should not the total he then needs be forty-nine rather than, as the text reads, forty-seven?[2] (2) What was the difference between Eudoxus' (or Callippus') idea of an astronomical model and that of Aristotle? In particular does that difference correspond to one between what we may call an instrumentalist view and a realist one?[3]

I do not deny that those questions are worth addressing. But there are other, prior, questions that are also important and rather more neglected. I wish to devote much of what I have to say to four in particular. (1) How much did Aristotle know about astronomy? (2) What did he need astronomy for? (3) How successful was he in fulfilling *that* aim? (4) What are we to make of the recurrent methodological disclaimers he makes in this field?

The complexity of the interrelations between these questions does not allow them to be dealt with sequentially, but, if I may anticipate, the conclusions I shall propose go like this.

---

[1] One modern discussion that suggests that the total number of spheres, direct plus retroactive, needed was not 55 (as at *Metaph.* 1074a11f.) but 61, is in Hanson 1973, p. 61.
[2] *Metaphysics* 1074a13f., already the topic of debate in Simplicius, *Commentary on Aristotle On the Heavens* (*Cael.* 503.10ff.)
[3] Aristotle's own account, with its unified interacting spheres transmitting motions inwards from the outermost heaven, is evidently realist: the question is, are we justified in assuming that the models of Eudoxus and Callippus were not also? That assumption has been challenged by Wright 1973–4.

(1) His knowledge is interestingly patchy. It is, in places, quite detailed, and some references could even be taken to suggest first-hand involvement, as an observer, and as a participant in theoretical debate. Yet there are also large gaps, and on important topics.

(2) Secondly, he needs astronomy primarily, I shall argue, for his teleology, especially to establish the orderliness of the heavens.

(3) Thirdly, he faces considerable difficulties in that enterprise (establishing order in the heavens). There are difficulties he knows about and discusses, but there are also some he knows about but skates over. His performance, in this area, is very similar to Plato's – despite the general contrast between his methodology and that of Plato. Indeed there are one or two places where it may appear that he is principally concerned just to score points against Plato.

(4) Fourthly, some of his methodological disclaimers do not carry conviction and even contain an element of bluff. How can a disclaimer be bluffing? I hope to make this clear.

One famous text in the *de Partibus Animalium* (I 5, 644b22ff.) gives us one answer to the question of what the study of the heavens is good for. There he contrasts the study of what is imperishable with the study of living things. We can learn more about the latter: but the objects the former studies are far superior. They are honourable and divine, even though the means of studying them are less, since there is altogether very little that is clear to perception to use as a basis from which one might investigate them and the things we yearn to know. Even so, he goes on at 644b31ff., 'even though our grasp of them is only slight, yet because they are so precious it brings us more pleasure than knowing everything in our region, just as a partial glimpse we chance to get of the beloved [in the plural, and of indeterminate gender] brings us more pleasure than an accurate vision of other things, however many or great they are'.

So his starting-points are clear. The heavenly bodies are divine and of supreme importance. At the same time he recognises that their study is difficult because of their distance and the lack of 'what is clear to perception'.

That is appropriately modest and reserved. But let us now ask what data he has or claims to have, where they come from, how far he attempted to check or add to them. How much, in fact, does he know about astronomy? On some occasions he conveys a very different and more confident impression about aspects of this subject than the guarded remarks of *de Partibus Animalium* I 5.

Three texts, especially, refer to what he evidently believes to have been established by observations carried out over a considerable period of time. *On the Heavens* 270b13ff. does so in very general terms. 'In the whole of past time, according to the records handed on from one generation to the next, no change has appeared either in the whole of the outermost heaven or in any one of its proper parts.'

At *On the Heavens* 292a7ff. he is more specific as to his sources. He has just referred to an occultation of Mars by the moon and he adds that similar observations about the other stars/planets have been made by the Egyptians and Babylonians, who have watched them 'for very many years' and to whom 'we owe many trustworthy grounds for belief (*pisteis*) about each one of them'.

In the *Meteorologica* 343b9ff., 28ff., he again refers to Egyptians for confirmation that some of the fixed stars have been seen with tails like those of comets, and that conjunctions take place both between pairs of planets and between planets and fixed stars.

In the last two texts cited, Aristotle does not just refer to the work of Egyptian and Babylonian observers, but also cites what 'we' have seen. In *On the Heavens* 292a3ff. there is an account of the occultation of Mars by the moon when it was half full: it approached Mars which was then blotted out by the dark side (i.e. the unilluminated edge of the moon) and then re-emerged on the bright side. At *Meteorologica* 343b11ff. he reports an observation of a tail to one of the stars in the Dog, though 'only a faint one. To those who stared directly at it, the light used to become dim, but to those who looked at it askance, it was brighter.' In the continuation of the same chapter, he announces 'we have ourselves seen' (*autoi heōrakamen*) Jupiter in conjunction with one of the stars in the Twins, which it occulted but without any comet appearing (*Mete.* 343b30ff.).

These are all first-person plural statements. That by itself certainly suggests contemporary Greek observations, even though not necessarily ones carried out by Aristotle personally. But the circumstantial report of the heightened sensitivity to the faint light of the tail of the star in the Dog when viewed askance, i.e. on the periphery of the retina, may suggest personal interrogation of those doing the observing, if not personal observation. So too may the emphatic 'ourselves' at *Mete.* 343b30ff., where Aristotle is at pains not just to report what was seen, but also his negative result. This conjunction of Jupiter with a star happened with no comet appearing, and so could be cited as evidence against Democritus' theory explaining comets as the conjunctions of planets (*Mete.* 342b27ff.).

Elsewhere there are impersonal references to what appeared or was observed. One of the more circumstantial comes at *Meteorologica* 345a1ff., the comet in the archonship of Nicomachus (341–340 BC). It appeared in the equinoctial circle for a few days: it did not rise in the West and it coincided with a wind at Corinth. That could have been observed by Aristotle himself. But elsewhere he reports sightings he is unlikely to have made personally. His favourite comet occurrence is that of the so-called 'great comet', which he refers to several times (*Mete.* 343b1ff., 18ff., 344b34ff.). But that has been dated to 373–372 BC when Aristotle was a lad of eleven or twelve at Stagira – and in no position to verify the reports of the earthquake that came with the comet in Achaea. The occasions when he refers to Greek records of comets include *Meteorologica* 343b4ff. citing one in the archonship of Euclees, son of Molon, at the winter solstice. But that was 427–426 BC, long before he was born.

What about the quality of Aristotle's observational data? In line with the usual standards of his day, those he cites, both his own and those from the records he draws on, suffer from two main limitations, to which I drew attention in Lloyd 1979, p. 180. These are that he has no precise dating system and no system of celestial coordinates, whether equatorial or ecliptic, by which to locate objects or events in the heavens. His dates are never more precise than the archon year plus the month. His locations are all by reference to other stars or to cardinal points, north, south, east, west.

Clearly those are considerable handicaps for anyone attempting a *determinate* account of the movements of the planets: that will be relevant later. More than that, the disadvantages extend also to the difficulty of checking existing reports and they affect the usefulness of any attempt to increase the data base – not that there are clear signs of a sustained effort on Aristotle's part to achieve either goal.

Aristotle often defers to the experts in astronomy, as is well known from the texts in Λ ch. 8 in which he does so, namely

1073b10ff. and 1074a16f., and from several in the *On the Heavens* which we shall be reviewing later. But I now want to look at some points where Aristotle did not need *either special* expertise in astronomy, *or* very exact data, to identify problems for his theories. Recall that he is absolutely convinced (1) that the heavens are unchanging, (2) that the movements of the planets are uniform and regular. Those doctrines are fundamental to his whole cosmology and to his metaphysics. The order of the entire cosmic system depends on there being order in the heavens. One might have thought, then, that he would be particularly careful to consider anything that might be deemed to constitute a *prima facie* objection to those doctrines.

Now it is true that some of the more embarrassing data, from Aristotle's point of view, only came to light some time after he was dead, and so it can hardly be held against him that he did not discuss them. However the list of those that might have given him pause, had he known about them, is considerable. They include: (1) the precession of the equinoxes, (2) novae and supernovae, (3) sunspots, (4) the 'face in the moon', (5) comets, (6) the colours of the planets, (7) variations in the brightness of the planets, (8) variations in the apparent angular diameter of the moon, (9) variations in the shapes and lengths of the retrograde arcs of the planets.

Knowledge of the first three of these post-dates Aristotle, to be sure. The precession of the equinoxes was discovered, as Ptolemy tells us (*Syntaxis* VII. 2) by Hipparchus in the second century BC. Hipparchus was also responsible, according to a report in Pliny (*Natural History* II. 95), for observing what Pliny calls a 'new and different star that came into existence in his own time. Its motion where it shone led him to wonder whether this happened at all often.' Quite what this object was – whether a comet, a nova or a supernova – is not clear from this report. But again, so far as data available to Aristotle go, they do not include any indubitable reference to novae or supernovae. Of course after Aristotle, those who shared his view as to the unchanging heavens, were certainly disinclined to see any such, that is to recognise them as new objects. But that is another question.[4]

---

[4] It is salutary to reflect on the contrast with ancient Chinese observations. The Chinese, who were not inclined to consider the heavens unchanging, report novae and supernovae far more readily than was done in the West: see Xi Zezong and Po Shujen 1966, Xi Zezong 1981.

The same also applies to sunspots, where again, for whatever combination of reasons, the Greco-Roman record is silent on phenomena that the ancient Chinese nevertheless picked up.[5] There is an interesting contrast, here, with the admittedly much more obvious lack of uniformity in the appearance of the surface of the moon. Aristotle himself refers to that at *On the Heavens* 290a26f., where he remarks that the moon always presents its 'so-called face' to the earth. Yet even though this might have suggested some corresponding variation in the physical constitution of the moon, the phenomenon was, at least, reassuringly stable, and as such could have been used to suggest constancy rather than inconstancy in the heavens.

With the fifth instance, listed above, namely comets, we have already seen that Aristotle has some specific data to hand, and certainly the phenomena under some interpretations conflicted with Aristotle's own theories. He refers to earlier views of comets that represented them as planets that appear only at great intervals (a theory ascribed to some Pythagoreans at *Meteorologica* 342b29ff.). At *Meteorologica* 343a22ff. he rejects that view principally on the grounds that the planets all appear in the circle of the zodiac, where indeed they undergo retrogradation, while many comets have been seen outside that circle. He further asserts that several comets have sometimes been visible at the same time and sometimes in conjunction with all five planets (343a25f., 30ff.).

Now the first argument must have seemed to carry some force. The band of the zodiac was defined by the greatest movement in latitude by any of the visible planets and moon. The Pythagoreans' reply would have had to be that the comets exhibited a greater deviation in latitude than any other planet, not that they were not planets. But all that Aristotle's final argument shows is that the comets cannot be identified with any of the visible planets, of which there are, he insists, just five. But that by itself does not show that a comet is not an *additional* planet.

Aristotle has, then, an alternative theory of comets to contend with, the implications of which, for his own system, would be severe. For if the theory of comets as planets were accepted, then until such time as their periodic returns had been established, their appearing 'at long intervals' would certainly seem to threaten the principle of the order and regularity of heavenly motion.

[5] See Xi Zezong 1981.

But how does Aristotle himself account for the phenomenon? In *Meteorologica* 1 7 he offers a possible account, claiming nevertheless that this is a sufficient demonstration concerning matters that are 'unclear to perception'.[6] His starting-point is that the outermost part of the terrestrial region, below the circular revolution of the heavens, is composed of a dry and hot exhalation: this and the air that is continuous with and below it are carried round by that circular revolution. When this ignites, that forms shooting stars. But when there is a 'fiery principle' that impinges upon a suitable condensation – a principle that is neither so strong as to cause a rapid conflagration, nor so weak as to be quickly extinguished – and when further it encounters a well-mixed exhalation from below, then a comet is produced.

That makes comets essentially *sublunary* phenomena, where, of course, change and variation are only to be expected. Yet he adds a further theory, *Mete.* 344a35ff., associating some comets more directly with the planets and even with the fixed stars, namely when the exhalation is constituted by their movement. Quite how this happens is not exactly clear. He compares the comet's tail with the haloes round the sun or moon, though he says that unlike the latter, where the colour is just a matter of reflection, the colour belongs to the tail itself. As in his explanation of the Milky Way, in the next chapter, 345a11ff., he takes it that, although the stars themselves are made of aither, their motions can somehow ignite the material in the sublunary region. But the conclusion of the whole discussion, at *Mete.* 346b10–15, has him repeating that all the phenomena he has been talking about, comets included, belong to the region round the earth that is continuous with the heavenly motions.

His handling of some other well-known data does not exactly inspire confidence. Take the differences in the colours of the planets. That is very obvious and was explicitly referred to by Plato, for example in the Myth of Er in the *Republic*, 616e f. Yet such differences could easily be accommodated by Plato, who had, after all, insisted (*Rep.* 530b) that it is absurd to think that the heavenly bodies, being visible and corporeal, are not subject to change. But for Aristotle, the planets, like the stars, are all made of the fifth element aither, and so how the colours they and indeed

---

[6] *Meteorologica* 344a5ff., cf. above ch. 1 p. 25.

the sun and moon themselves appear to have can vary needs, one might have thought, some explanation – a point eventually urged against Aristotle by, for example, Philoponus.⁷ Yet no explanation for the different appearances of the planets is forthcoming: indeed Aristotle does not mention those differences at all.

What he does mention, on several occasions, is their not twinkling, *ou stilbousin*. One prominent text in which he does so is that in the *Posterior Analytics* 78a30ff., where he points out that although non-twinkling and nearness are convertible, it is nearness that explains non-twinkling and not vice versa. The syllogism that uses nearness as the middle term explains the reasoned fact, the *dioti*, not merely the fact. But our admiration for the clarity of the logical point Aristotle makes should not lead us to ignore what a strange example he uses. I do not just mean that 'non-twinkling' is a negative term, and 'near' a quite imprecise one: I have criticised the example on those grounds before.⁸ The empirical basis for Aristotle's point is also strange.

To start with, it seems to conflict with one common Greek name for the planet Mercury, namely *Stilbōn*, the twinkler.⁹ But then there is a further point. Aristotle was evidently interested in how the planets shine, the quality of their light, though, to be sure, he did not have to keep checking that they were not twinkling. But then he ought to have noticed (one might think) the variations in their brightness, particularly noticeable in the cases of Venus and Jupiter, even if less so in such a case as Saturn. In some instances the brightness varies by several orders of magnitude, depending (1) on their distance from the earth, and (2), in the case of the lower planets, also on their phase.

Yet again there is no mention of this in Aristotle. It was, of course, one of the points on which the concentric spheres hypothesis eventually foundered, to be replaced by epicycles and eccentrics. With concentric spheres, the planet, located on the innermost sphere, is at a constant distance from the earth, whereas variations in that distance can easily be accommodated on either the eccentric or the epicyclic model.

---

[7] Philoponus' lost work, *Against Aristotle On the Eternity of the World*, is quoted by Simplicius, *Cael*. 88.31ff., cf. Wildberg 1988, pp. 181ff.
[8] See ch. 1 pp. 13f. and 33.
[9] This is what Mercury is called in the *Ars Eudoxi*, 5.10, and cf. the pseudo-Aristotelian *de Mundo* 392a26.

It might be urged, in Aristotle's defence, that the variation in the brightness of the planets is not a very prominent phenomenon, and that such extraneous factors as the atmospheric conditions of observation might have been thought sufficient to account for such variations as he recognised. However, an extended passage in Simplicius suggests that already in the late fourth century BC the variations in the brightness of the planets were known and had been taken to imply that their distances from the earth vary.[10] Simplicius reports that both Eudoxus and Callippus failed to deal with the problem this posed for their model, and it would certainly appear possible for Aristotle to have registered the phenomenon.

A similar difficulty – with data that seem to suggest that the distances of the non-fixed stars vary – arises in the case of the moon. The variations in its apparent diameter could hardly be observed directly with the techniques available in the fourth century BC. But in total eclipses of the sun the relative distances of the sun and moon do become rather strikingly visible. When the moon is close to apogee, there is the phenomenon known as an annular eclipse, with the rim of the sun still visible around the dark circle that corresponds to the interposition of the moon.[11] At perigee, by contrast, the moon more nearly causes total obscurity of the sun's disk.

By now it might be thought that I have been pressing points that are quite unfair, either because the phenomena in question are not obvious or because they were peripheral to Aristotle's main concerns. But neither defence will work with the next problem, that of the speeds of the non-fixed stars and their stations and retrogradations, since it was precisely there that Eudoxus' concentric theory was supposed to give a good first-stage approximation to a solution which Aristotle himself felt confident enough about to adapt for his own metaphysical purposes.

But just how confident was Aristotle? We have not only *Metaphysics* Λ 8 to consider, but other texts too that broach the problems to do with the complex movements of the planets. There are well-known doubts about the reliability of Simplicius' testimony, reporting Sosigenes, who may in turn be drawing on Eudemus, to

---

[10] *Cael.* 504.17ff., where Simplicius cites Sosigenes as his source.
[11] This is referred to, for example, by Sosigenes, according to Simplicius, *Cael.* 504.25–505.19, and cf. Cleomedes, *de Motu Circulari* 190.17ff.

the effect that Plato set astronomers the task of accounting for the apparently irregular motions of the planets by combinations of uniform, regular, that is, circular, movements. Yet whether or not that report is correct, Plato himself expresses a keen interest in the problems of planetary motion. Even though he draws back from giving a detailed account in the *Timaeus* (40d2ff.), he identifies the Circles of the Same and of the Other that provide, in principle, for the two main periodicities of each of the planets, sun and moon, that is both their daily westerly movement with the sphere of the fixed stars, and their varying easterly motion along the ecliptic. In the *Laws*, moreover (822a), he protests vigorously against the idea that the planets 'wander' and even claims that each moves with a *single* circular movement.

At the time when Aristotle entered the Academy, it is likely enough that how to account for the movements of the planets was seen as the key question in astronomy, and its cosmological implications for the notion of order in the universe in general would not have been lost on anyone either. Aristotle's own first forays in the field, to judge from texts outside *Metaphysics* Λ 8, were appreciably less detailed than what we have there. He does not have anything to say about how concentric spheres may be combined in the *Physics*, though he does recognise that some of the heavenly bodies move with a complex motion (e.g. 259b29ff.). The difference between the primary, westerly, motion of the sphere of the fixed stars, and the secondary, easterly, movements of the planets, sun and moon along the ecliptic, is recognised, and the latter motion, in the case of the sun, provides the basis of Aristotle's explanation of change in the sublunary region (cf. *On the Heavens* 286b1ff., *On Generation and Corruption* 336a15ff.).

One minor problem that might be thought to arise is how these two types of movement can be squared with the doctrine that there is a fundamental contrast between rectilinear (sublunary) motion and circular (celestial) motion in that the former can, the latter cannot, have an opposite. Rectilinear motion is either to or from the centre of the universe ('up' or 'down'). Similarly clockwise and counterclockwise movement might be thought to form a perfectly good pair of opposites, so far as circular motion is concerned. But Aristotle will have none of that. In *On the Heavens* I. 4, 270b32ff., he argues that opposite in the case of motion means from or to an opposite point. Clockwise and counterclockwise movement from the same point on a circle do not meet that requirement. No more will points at opposite ends of a single diameter of a circle provide proper opposite *archai*, starting-points, for motion.

Nevertheless Aristotle does recognise a number of difficulties. Three of the main ones in *On the Heavens* are: (1) why do the heavens revolve in one direction rather than in the other (II 5)? (2) Why do the planets have different speeds (II 10)? (3) Why do the non-fixed stars have complex movements, while the fixed stars have a simple one (II 12)?

In all three cases, Aristotle qualifies or hedges his account. (1) In II 5, 287b28ff., he says: 'perhaps to try to make declarations about some things, indeed about everything, omitting nothing, might well seem a sign either of much simplemindedness or of much zeal'. However he goes on, if we aim at a merely human account, that criticism is not fair. If (*men*) someone hits upon more exact necessities, then we should be grateful to the discoverers; but as it is (*de*), we must state what seems to be the case.

(2) In II 10, 291a29ff., we are told that the order and distances of the planets should be investigated on the basis of astronomy (*ek tōn peri astrologian*), where they are adequately (*hikanōs*) discussed. But it happens that the outermost motion of the heavens is simple and the swiftest, those of the inner spheres slower and composite. So it is reasonable (*eulogon*) that the planet that is closest to the outermost revolution (viz. Saturn), completes its own motion in the longest time, for it is subject to the strongest influence from the outermost revolution: while furthest away from that revolution, the motion is least affected, and swiftest. The speeds of the intermediate motions vary with their distances, 'as indeed the mathematicians show'.

(3) In II 12, 291b24ff., the two problems of the chapter are to be investigated despite their difficulty, though the eagerness to do so should be considered a sign not of rashness but of modesty, if, out of a thirst for philosophy, one is content with little by way of solution to the greatest difficulties. However it is worth while seeking to extend our understanding (292a14ff.), even though we have very little to start from and are very far away from the occurrences we are trying to investigate. Nevertheless (*homōs de*) if we investigate on the basis of those starting-points, the present difficulty would not seem to be unaccountable.

So in each chapter we have a warning, as to the difficulty, or as to the need to consult the mathematicians, but in each case that is followed by an expression that justifies, at least up to a point, the account that is then attempted. But it is not just that there is this balance between 'on the one hand' and 'on the other'. The relationship between the quality of his tentativeness and the grounds for his positive assertions is interesting.

Thus it is surely significant that both on the problem of why the heavens revolve in one direction rather than in the other – in II 5 – and on the difficulty of the complexities of the movements of the non-fixed stars – in II. 12 – his positive speculations invoke teleology. Nature produces the best result possible, he says in II 5, 288a2ff., and so the movement of the heavens is forward not backward, to the right and rightwards, as he had insisted in II 2, as animal movement should be (a doctrine that leads him to the conclusion that the north celestial pole is the lower of the two, 285b22ff.). The zoological analogy is pursued even further in II 12. There the planets are compared with humans, in that both have varied actions with varied goals. The other animals have less variety of action, and the plants just one. Similarly, he claims, the sun and moon have fewer motions than some of the planets (291b35f.) and the earth no movement at all. 'In this way, he concludes, 293a2ff., nature balances things and introduces a certain order, assigning many bodies to the one motion [that of the fixed stars] and to the one body many motions [in the case of the non-fixed stars].'

What emboldens Aristotle, on both occasions, is the belief that the heavenly bodies, and the heavens themselves, are alive. Indeed as living creatures they are far superior to humans, who are, in turn, higher than the other animals, in the basic hierarchical and teleological schema that articulates the whole of Aristotle's world-view.

Then on the side of the reservations expressed, II 10, on the order, distances and speeds of the planets, is particularly cool. The order and distances should be investigated from astronomy, where they are adequately discussed. 'Adequately' suggests Aristotle was satisfied with what the astronomers had had to say. Certainly there was agreement about the order of the outer planets, Saturn, Jupiter, Mars. Plato continues with Venus, Mercury, Sun and Moon, but that was not so straightforward. At least some later Greek astronomers were to adopt the so-called Chaldaean order, with the Sun in the middle of the system.

As to their speeds, II 10 adopts the usual view that the outermost revolution is simple and swiftest, whereas those of the inner planets are slower and composite. But having contrasted Saturn's period of return (slowest) with that of the moon (swiftest), Aristotle puts it, as we noted, that the speeds of the intermediate motions vary with their distances, again citing the mathematicians as his authority. But that looks like a gross oversimplification. If we take it that sidereal revolutions are meant, the figures that Eudoxus is reported to have used for the outer three planets work well enough, that is thirty years for Saturn, twelve for Jupiter and two for Mars.[12] But – as Plato already knew (*Timaeus* 36d5) – Venus and Mercury have the same sidereal revolution, namely one year, as the sun. If we take it that it is not the sidereal revolution that is meant, but actual linear speeds, which will vary with the distance of the heavenly body, the trouble there was that the astronomers of Aristotle's day (and indeed much later) were in no position to arrive at independent figures for the distances, let alone to 'show' them. It thus appears that Aristotle's remark, that the speeds vary in proportion to the distances, is a very loose one, and that in turn suggests that the reference to what the 'mathematicians' 'show' is partly just a matter of hand-waving.

A further problem of a similar nature arises in connection with II 12. There the sun and moon are said to move with fewer motions than *some* of the planets (291b35: *enia* at 292a1). That was certainly not Callippus' view, for he had five spheres each for the sun and moon, five for three of the planets (Mercury, Venus, Mars) and four for the other two (Jupiter, Saturn). But it does not tally with Eudoxus' theory either, for he had three spheres each for the sun and moon, and four each for *each* of the planets. So his view would be that the sun and moon move with fewer motions than *all* of the planets, not just some. Nor does what Aristotle says tally with anything in Plato, who has Circles of the Same and Other but no concentric spheres at all. Nor finally can we save the text of 292a1 by reference to the solution eventually preferred by Aristotle, in *Metaphysics* Λ 8, with its introduction of retroactive spheres, for the 55-sphere version gives the sun more motions (nine) than two of the planets (Jupiter, Saturn) and as many as the other three, while

---

[12] Simplicius, *Cael.* 495.26ff., gives figures that correspond to, and may have been derived from, Babylonian values.

Table of the number of spheres required for the planets, sun and moon in the models of Eudoxus, Callippus and Aristotle

	Moon	Sun	Mercury	Venus	Mars	Jupiter	Saturn	Total
Eudoxus	3^a	3^a	4	4^b	4^b	4	4	26
Callippus	5^c	5^d	5^c	5^e	5^e	4	4	33
Aristotle								
On the Heavens II. 12: the moon and sun have fewer motions than *some* of the planets, 292a1								
Metaph. Λ. 8								
Mark I (Callippus + retroactive spheres, 1073b38ff.)								
	5	5+4	5+4	5+4	5+4	4+3	4+3	55
^{f} Mark IIA (Callippus modified + retroactive spheres, 1074a12ff.)								
	3	3	5+4	5+4	5+4	4+3	4+3	47
^{f} Mark IIB (IIA but with retroactive spheres for the sun reintroduced)								
	3	3+2	5+4	5+4	5+4	4+3	4+3	49

^{a} Neither moon nor sun retrogrades
^{b} On given parameters (from Simplicius) Venus and Mars cannot retrograde
^{c} In the later eccentric-epicyclic system of Ptolemy, the moon and Mercury are recognised as the two most complex motions
^{d} Callippus' extra spheres for the sun were, according to Eudemus in Simplicius, to explain the inequality of the seasons
^{e} Conjecture: Callippus' extra spheres for Venus and Mars were to secure retrogradation
^{f} The Mark II version, in Aristotle, reverts to Eudoxus on both moon and sun

the 47-sphere version gives Sun and Moon fewer motions than *all* the planets, as indeed also does the proposed 49-sphere amendment. The table sets out the figures in each case.

That may seem a harsh criticism to make of what many would wish to excuse as a work of Aristotle's youth. But whatever the explanation, or excuse, we have to register that there are few signs, in these three chapters of *On the Heavens*, of a grip on technical detail. His primary interest or concern is teleology, and his main resource, when encountering difficulties, is the appeal to the belief that the heavenly bodies are alive. True, his solutions are proposed tentatively, and he defers to what he claims the astronomers had shown, leaving open the possibility, in II 5, that more work might alter the picture. Yet those allusions to the authorities would have carried more conviction if they had been backed up by more definite indications that he had followed their technical discussions through with greater attention to detail. In the absence of any such indication, the references to their work and to what they had shown risk seeming pure eye-wash.

Yet *Metaphysics* Λ 8, at least (to turn to our main text at last), does surely show quite a grasp of the intricacies of contemporary astronomical model-building. Our next task is to try to analyse the relationship between the signs of tentativeness, and the positive proposals, in this, at once the most important and the most difficult of Aristotle's forays into astronomy.

The discussion is framed by two remarks. First at 1073b3ff. he says that for the number of the motions one must turn to astronomy. As for how many they happen to be, he continues at b10ff., to give some idea of the subject he will report what some of the mathematicians say, in order to have a definite number for the mind to grasp. The possibility that further research will produce conclusions that conflict with his present account is mentioned: in which case, as he puts it, b16f., one should love both parties, but believe the more accurate.

Then again at the end of the account, at 1074a14ff., he says: let this then be the number of the spheres, so it is reasonable to suppose that the unmoved substances and principles are as many. But as to a statement of necessity, that may be left to more powerful thinkers.

There can be no question of taking Aristotle's astronomical account as anything other than a very tentative one: certainly it is presented as in no sense a definitive one. At the same time he does claim to make a personal contribution in this area. In practice, it was he, of course, who introduced the notion of retroactive spheres, 1073b38ff., whose function, in his own words, was to restore to the same position the first sphere of the heavenly body immediately below. Moreover in principle, in his remark on the possibilities for further research, 1073b13ff., he refers not just to learning from (other) investigators, but also to investigating himself.[13]

The frame of his account contains two further remarks that are of some interest. When he embarks on his discussion of the number of the movements, he begins (1073b8ff.): that the movements are more than the bodies moved is clear even to those with a

---

[13] It is clear who the *autous* of 1073b14 must be, for they are the subject of the two verbs, *philein* and *peithesthai* at b16.

modicum of experience (*metriōs hēmmenois*), for each of the non-fixed stars is moved by more than one motion. That may seem a surprising remark for him to make. He does not just refer to his own discussion of the complexity of planetary motion in, for example, *On the Heavens*. Nor does he just assume that these motions are complex. He points it out. Anyone with even a little grasp of the subject knows that. Yet one person who had *denied* it was Plato, who in that famous, but problematic, passage in the *Laws*, 822a, already cited (p. 169 above), had claimed, as regards the planets, that each of them moves on a *single* circular track, though appearing to move on many. While Aristotle does not refer to Plato, readers of the *Laws*, I suggest, would not fail to pick up the implied criticism.

Nor is that the only point where Plato may be in Aristotle's mind. He goes on, as noted, at 1073b11ff., to report what the mathematicians say, in order to have a definite number, a *plēthos hōrismenon*, for the mind to grasp. What Plato had said in the *Timaeus* 39d, in striking contrast, indeed, to *Laws* 822a, was that the wanderings of the planets are of 'intractable number' (*plēthei amēchanōi chrōmenas*) and 'wondrously complexified' (*pepoikilmenas thaumastōs*).

Where one text in Plato had asserted that each planet has a single, circular, path, and another had gone to the opposite extreme of saying that their wanderings are of amazing complexity, Aristotle occupies the space in between, denying both simplicity, and untold complexity, in favour of a determinate number in each case. Again there is no citation of the *Timaeus*, any more than there was of the *Laws*. Aristotle's introductory remarks can be read – indeed usually are read – without reference to a Platonic background of any kind. Yet it would seem to me that readers of Plato would not have missed the point that in those remarks Aristotle distances himself very effectively from *both* the main texts where Plato had broached the question of planetary motions.

Aristotle needs his determinate astronomical account both to secure the orderliness of the heavens and to do better than Plato. But what can be said about Aristotle's own engagement in the subject? His personal contribution is the introduction of retroactive spheres which unify the whole system and imply the transmission of movement from the outermost sphere down to the moon and then on, in ways not discussed here, to the sublunary region itself.

But that contribution is motivated by cosmological, not purely astronomical, reasons. So far as the astronomy of the direct spheres goes, he is not sure which model to use. Eudoxus had attempted one solution, which Callippus had then modified with the introduction of further spheres. Whether Aristotle's own retroactive spheres achieve the function for which they were introduced, and whether they do so in the most economical fashion, are, as I noted at the outset, disputed questions. But leaving them aside, the issue I now want to focus on is what Aristotle's treatment of the two models he cites can tell us about his grasp of contemporary astronomy.

Our answers to that question are bound to be speculative, since so too is any reconstruction of the models of Eudoxus and Callippus.[14] Fortunately we do not need to enter into all the details of those problems to make some progress on the issue I raised. At least it is clear that Aristotle first applies his retroactive spheres to the Callippan system, and then to a modified version of that system, where he omits (so he says, 1074a12ff.) the extra spheres for the moon and sun, to get to the final total of (as our texts have it) forty-seven spheres. Why he should have proceeded, at that point, in precisely the way he does, is worth asking, even though some parts of the answer are bound to be uncertain.

The Eudoxan model, it is generally agreed, provides well enough first for the daily westerly motion of the planet, and then also for its generally easterly motion along the ecliptic. Moreover the great strength of the account was that, in principle, it could describe retrogradation. This it did by means of the two innermost spheres of each planet, which together produce the three-dimensional figure of eight known as the hippopede. Thus far the structure of the model is clear at least in qualitative terms, and some – incomplete – quantitative parameters are reported by Simplicius.

Thereafter, however, the problems begin to accumulate, both for our reconstruction, and for the system insofar as it can be reconstructed. First the very incompleteness of the data in our sources thwarts a detailed interpretation of the system. Indeed whether it was a *fully* quantitative model is very doubtful. A second difficulty that increases those doubts relates to some of the data that are reported by Simplicius. The parameters he supplies for Mars and Venus have the disastrous consequence that in neither

---

[14] The secondary literature on this topic is immense and includes Heath 1913, Dicks 1970, Maula 1974 and especially Neugebauer 1975.

case, on those values, is retrogradation possible.[15] Then a third problem, explicitly noticed by Simplicius, is that Eudoxus' model failed to account for the inequality of the seasons, although that had been well known since the work of Meton and Euctemon around 430 BC and was indeed recognised by Eudoxus himself.[16] Although he postulated three spheres for the sun, the third sphere, according to our reports, was designed to account for a – purely imaginary – latitudinal deviation of the sun from the path of the ecliptic.[17]

Now Callippus, again according to Simplicius, introduced two extra spheres for the sun in part in order to meet the third of the difficulties just mentioned, namely the inequality of the seasons. Quite how he deployed the two extra spheres for the moon, and the one extra for Mercury, Venus and Mars, that he also postulated, are matters of pure conjecture. But of course the moon's motions are very complex, and so too are those of Mercury: and in the case of Mars and Venus, Eudoxus' model, as I have noted, seems to have failed to yield retrogradation at all. No doubt Callippus' general motivation was to give a better fit to the known phenomena: but we can gain some inkling as to the particular reasons for his modifying particular parts of Eudoxus' model. There is indirect confirmation of that in that, in the case of the two outermost planets, Saturn and Jupiter, where that model may well have worked best, Callippus left the number of spheres needed unchanged.

While every allowance has to be made for the conjectural nature of that – partial – reconstruction, we may next ask what light it would throw on the way Aristotle reacted to these two models. As noted, he first applies his own retroactive spheres to unmodified Callippus, 1073b38ff., 1074a6ff. That suggests that he thought Callippus superior to Eudoxus. But he then considers not adding the extra Callippan spheres for the sun and moon, 1074a12ff.

How far can we make sense of that? In general, on the basis of the partial reconstruction offered above, it looks as if Callippus' new spheres were introduced to try to meet some of the major difficulties apparent in Eudoxus' model. On this argument it is not at

---

[15] See, for example, Heath 1913, p. 211, Maula 1974, pp. 73ff., Neugebauer 1975.
[16] This is clear from the figures for the seasons given in the *Ars Eudoxi* 22.1–23.14.
[17] Simplicius, *Cael.* 493.11ff.

all surprising that Aristotle would have in general favoured Callippus. Thus it seems that Aristotle was satisfied that the extra spheres introduced for the lower three planets, Mars, Venus and Mercury were necessary. Again he starts off following Callippus on the moon and sun, and again it is not difficult to conjecture ways in which Callippus might be thought to have the edge over Eudoxus. Yet if we follow Simplicius here, and take it that Callippus' extra two spheres for the sun were an attempt to deal with the problem posed by the inequality of the seasons, we are then faced with the puzzle of why Aristotle should consider reverting to the Eudoxan model, when at the end he contemplates *not* adding the extra Callippan spheres for the sun and moon.

This final remark on the sun and moon shakes one's confidence. It may be that the whole reconstruction is at fault, though what I have presented here is common ground to most interpretations and has the advantage of sticking close to what is actually reported by Aristotle and Simplicius themselves. It may also be that all that Aristotle wishes to do, in that final remark, is to underline his own hesitancy: all he is really concerned with is the connection between the number of the spheres and the number of the unmoved movers, and provided he has some determinate number for the former, he has the answer to the number of the latter. Yet he expresses that hesitancy in a context and in a manner that – if the reconstruction is sound – may suggest he is seriously out of his depth. Let me explain.

The inequality of the seasons was both well known and a major problem for the Eudoxan model. Yet *either* Aristotle remained ignorant of it *or* he thought it no problem – or not one, at least, that extra spheres could solve. The first possibility cannot be ruled out entirely. Aristotle does not refer to the fact directly (any more than does Plato), though it would be amazing if he had heard nothing of the work of Meton and Euctemon, in the 430s, let alone that of Eudoxus himself, on this question.

But if he did know about it, the surprise then becomes how he could contemplate the possibility that, after all, no extra spheres might be needed for the sun. True, we do not have the details of the Callippan solution to the problem. But within the framework of the concentric spheres hypothesis (the only framework in view) the only way of meeting this major difficulty concerning the sun's movement was by employing extra spheres.

Heavenly aberrations

The possibility of not needing extra direct spheres for the sun and moon is expressed very tentatively at *Metaphysics* 1074a12ff., in line with the tentativeness of the whole account, and yet it is surprising. But having registered my surprise, I have to confess that I have no explanation to offer. It is not as if Aristotle is simply repeating the view expressed in *On the Heavens* II 12, where he remarked that the sun and moon have fewer motions than *some* of the planets (291b35f., see above pp. 171f.). That will not do, since on the modified Callippan model the sun and moon both have fewer spheres (direct, or direct plus retrograde) than *all* the planets.[18]

Of course neither the sun nor the moon goes retrograde, and so from that point of view, their motions ought to be simpler than those of the planets, for they all do and will, on the Eudoxan model, need hippopedes to do so. Yet *if* that *were* Aristotle's thought, it would be the final nail in the coffin of his reputation *as an astronomer*. For it would be as if he just focussed on the single fact of non-retrogradation, and ignored all the other complexities, both of the moon's movements and of the sun's.

Aristotle would have us believe he is an amateur of astronomy. There are those glimpses of the beloved that mean so much to him. Yet he might have looked a little harder, and he could have benefited from more sustained discussions with the likes of Eudoxus and Callippus. It is all very well to say that both should be loved, but one should believe the more accurate. But he appears not to have engaged closely enough with the subject to make a real contribution to contemporary astronomical debate, and maybe he was not even close enough to judge who indeed was the 'more accurate'.

What one would have expected from someone more closely involved in that debate is a sense of where the important issues lay, in such matters as how to account for the inequality of the seasons, how to get the best fit for the retrograde arcs of each of the planets, how to deal with the Achilles' heel of concentric

---

[18] The moon and sun each had three spheres, in Eudoxus' model, and the sun, it is thought, needed two retroactive spheres (though in that case the total number on the modified Callippan model should have been 49, not 47). But for Mercury, Venus and Mars, the model called for five direct plus four retroactive each, and for Jupiter and Saturn four direct plus three retroactive. The table on p. 173 sets out the figures on each model for ease of reference.

theory, the apparent variation in the distances of the non-fixed stars, as judged by their varying brightness and apparent diameters. His overriding concern, in *Metaphysics* Λ 8, is with the two questions, of the number of the spheres needed, and their unity: the first has to be answered, however provisionally, to answer the question of the number of the Unmoved Movers, and the second is essential for his account of the transmission of motion from the outermost sphere to the sublunary region.

But by the same token he exhibits little interest in the details of how well the various models on offer do the astronomical job of work they were designed for, nor with indicating where the major remaining problems (not to say the perhaps insuperable difficulties) lay. He mentions differences between Eudoxus and Callippus but does not analyse the strengths and weaknesses of their rival models, notably, for example, how good a match they could give for the observed retrogradations of the planets, which vary both as between one planet and another and as between one retrogradation of the same planet and another. A practising astronomer would have felt the need to address those questions, both with a mathematical analysis of the resolution of complex movements into circular components, and with an empirical review of the data available.

By those criteria, Aristotle was no astronomer at all – nor does he present himself as a direct rival to Eudoxus and Callippus, even though he does refer to his own personal contribution to the investigation, 1073b13f. Rather, what he needed his astronomical account for (to turn to the second of the questions I stated at the outset) is its role in his cosmology. It is his own confidence in his teleology that sustains him in the face of the difficulties mentioned in *On the Heavens*. He shares Plato's strategic interests in this area, and the comparison with Plato is apposite, if he is in the background both when Aristotle protests, in *Metaphysics* 1073b8ff., that anyone who knows anything knows that the planets have several motions, and when he goes on to insist that each has a determinate (and by implication not an intractable) number of spheres.

My third initial question was how successful he was in fulfilling his aim. We can be pretty sure he was confident that astronomy could be used to recommend the thesis that the heavens are orderly, eternal and unchanging. Though he expresses doubts about such matters as the number of the spheres, he never wavers on that thesis. Yet we have to marvel. Plato, after all, had said that it would be absurd to believe that what is corporeal and visible is unchanging. Again Aristotle knew of the view that the comets are planets, though he rejects it. Again there are differences in the appearances of the heavenly bodies, in their colours, for instance, that might be thought difficult to square with the belief that they are all made of perfect aither. Even on Aristotle's own picture, where the movements of the planets are the resultant of spheres moving regularly but at different angles of inclination and with different speeds, those postulated differences remain unaccounted for. It may be acceptable to abandon some tricky problems in the sublunary region on the grounds that it is subject to imperfections of many different kinds. But differentiation in the heavens calls more loudly for some explanation. There is a striking contrast here between *On the Heavens* and the *Metaphysics*. When, in the former, he is arguing for the thesis of the perfection of aither, its eternal, unchanging, circular motion, the full complexity of his (eventual) planetary model does not obtrude:[19] when it comes to occupy his full attention, in the *Metaphysics*, it is, one might say, too late, because the perfection of aither is now an unshakeable belief.

But then a further even more striking contrast may be remarked between the expectations generated by his philosophy of science and his actual delivery in the matter of the strictest style of demonstration.[20] That style is often illustrated with mathematical examples. But arithmetic and geometry, as he puts it at *Metaphysics* 1073b6ff., do not deal with substance as such. So astronomy, as the same passage has it, b4ff., as dealing with perceptible, but eternal substance, has pride of place as the mathematical study closest to philosophy. But that might lead one to expect that it would yield plenty of first-rate examples of the most rigorous demonstrations. True, there are a few astronomical illustrations that play prominent roles in the *Posterior Analytics*, relating, for instance, to the causes of eclipses, and to the phases of the moon.[21] Yet the total haul is meagre. When it comes to the discussion of planetary motion, necessity is firmly disclaimed, that is left for other, more

---

[19] This remains the case even though there are general references to the complexities of planetary motion especially in the two chapters (II 10 and 12) discussed above (pp. 170f.).

[20] Cf. above ch. 1.

[21] e.g. *Posterior Analytics* 78b4ff., 90a7ff., 93a29ff., 95a14ff., 98a37ff. Cf. above, p. 167, on 78a30ff. on the cause of the non-twinkling of the planets.

powerful thinkers (*Metaph.* 1074a16f.),[22] nor are causal explanations – explanations of the reasoned fact, the *dioti* – in evidence. When it comes to comets, we are reduced to that looser style of demonstration that brings the phenomenon back to what is possible (*Mete.* 344a5ff.).

So we come, in conclusion, to the evaluation of those prominently placed disclaimers in his astronomical discussions. Both the similarities with, and the differences from, the methodology and the practice of the zoology are alike suggestive. In both zoology and astronomy we often find him stressing the difficulty of the subject-matter, including the difficulty of observation, the need for further research, the tentativeness of his own provisional conclusions. The difference is that in the zoology he speaks as a totally engaged researcher, one with as much first-hand experience of the problems he discusses as any of his contemporaries.

That can hardly be said of his astronomy. There too we have frequent expressions of tentativeness and he several times defers to what the mathematicians say. From one point of view that is reassuring, for he is certainly avoiding dogmatism on a number of points where he rightly feels it to be unwarranted. But from another point of view, those disclaimers do not cut much ice. It is all very well for Aristotle to say, on the problems of planetary motion, that he will love both parties, but believe the more accurate: but our analysis of the revised Callippan model mentioned at the end of his account, 1074a12ff., suggested that he may be seriously out of his depth, in a way that shakes one's confidence in his ability, indeed, to judge whose model was more adequate. At that point we might object that, even if no practising astronomer himself, he owed it to his audience to have done more homework in the field, if he wanted (as he did) to make use of some of its results. Again, it is all very well for him to say, in *On the Heavens* II 10, that the mathematicians' account of the distances of the planets was adequate and that they had shown certain facts concerning the relative speeds of the planets varying with their distances: but on close examination those comments, too, ring rather

---

[22] Cf. the reference to the possibility of more exact necessities being discovered, at *On the Heavens* 287b34ff. (above p. 170). Yet at the very beginning of *Metaphysics* Λ. 8, the Platonists are criticised for not speaking with demonstrative seriousness (*meta spoudēs apodeiktikēs*) on the question of the determination of numbers by the decad, represented as their solution to the general problem of the number of unmoved principles (1073a14ff. and 22).

hollow, for they reflect only at best a superficial grasp of the issues. In these circumstances his methodologically correct remarks sometimes appear as no more than window-dressing, or, as I put it, mere hand-waving, the gestures, one might say, of a star-struck lover.

CHAPTER 9

The idea of nature in the Politics[1]

It is well known that in a number of contexts, especially but not exclusively in the first book of the *Politics*, Aristotle invokes the concept of nature or the category of the natural in his discussion of problems in political theory: we would, no doubt, find the fact much more surprising, even shocking, if it were not so familiar. The most famous example, without a doubt, is the repeated statement that man is by nature, *phusei*, a *politikon zoon*. That slavery is a natural institution is a second, notorious, instance. A third is the distinction, in *Politics* I 1256a1ff. especially, between those features of household management, *oikonomikē*, that are according to nature, *kata phusin*, and those that are not; and other examples can readily be given. To the student of later European political theory it may look as if Aristotle is straightforwardly guilty of a breach in the principle most forcefully stated by Hume, to the effect that normative statements cannot be deduced from descriptive ones – that you cannot derive an 'ought' from an 'is'. Aristotle cannot exactly be accused of not having paid sufficient attention to Hume: though that should not deter us from using arguments from him and others in our assessment of the strength of Aristotelian positions. But we do not have to turn to later philosophers in order to identify a puzzle. How, one might ask, can that appeal to nature in the middle of an analysis of social and political issues be reconciled with the Greeks' own firm distinction between *phusis*

[1] This chapter represents a revised version of the original English text that I used as the basis of a presentation to Professor Aubenque's seminar on 10 April 1987, subsequently published as 'L'idée de la nature dans la *Politique* d'Aristote' in *Aristote politique*, ed. P. Aubenque (Paris, 1993), pp. 135–59. As I explained in that original publication, my aim was to stimulate discussion, rather than to present an exhaustive analysis of the issues. In this revision I have made some additions to bring the bibliography up to date, but I make no claims to comprehensiveness in that regard.

and *nomos* – a distinction used, of course, by among others Aristotle himself?

I shall divide my discussion into three uneven parts. The first – which constitutes the bulk of this chapter – will be devoted to commenting on the main contexts in which nature is invoked in the *Politics*. Are we dealing with mere assertion, or are there arguments that Aristotle could or does use in support of his positions, and if so how good are those arguments? This will lead me to consider, secondly, certain more general issues in his concept of *phusis* as a whole and to a comparison between its use in the *Politics* and elsewhere in the Aristotelian Corpus. Thirdly we must attempt to evaluate Aristotle's use of the category of the natural in political theory, particularly in the light of the charge that he has breached Hume's principle or otherwise made an obvious or fundamental category mistake: have we ourselves anything to learn from a study of *his* way of proceeding in this field of thought?

First, then, to some of the principal contexts in which nature is invoked at the heart of Aristotle's socio-political analysis. That man is by nature a political animal is, of course, represented by Aristotle as the conclusion of a piece of analytic reasoning. All animals (and indeed plants) naturally desire to reproduce (*Pol.* 1252a28ff.). Like other animals, male and female humans naturally form associations for reproductive purposes: so that basic mode of association, the household, *oikia*, is, we may say, founded on biological needs. But the household is not self-sufficient: to fulfil more than their merely daily, ephemeral, needs, several households agglomerate, first into villages and then into more perfect or complete associations, a process that culminates in the *polis*, which *comes to be* for the sake of life, but *is* for the sake of the good life (1252b28ff.). So every *polis* is by nature, *phusei*, given that the first associations or communities also are: for it is their end, *telos*, and nature is an end.

Now as is well known, many others before Aristotle had attempted some would-be historical account of the origins of civilisation, of the transition from primitivism to culture, and to that extent in those opening pages of the *Politics* Aristotle may be seen as developing traditional themes. But unlike some others who had, in more or less superficial fashion, used the *topos* of the similarities and differences between humans and the other animals in an account of the origins of human social organisation,[2] Aristotle was, of course, not just an acute political theorist but also an eminent zoologist and sociobiologist. I leave to one side the question of the precise chronology of his various zoological and political works, for I think it is sufficiently clear that – however far his particular researches in the one or the other domain had progressed at different stages during his career – Aristotle was *always* keenly interested in both domains, that is throughout his adult life. Sociobiology is indeed an appropriate enough term for those sections of the *Historia Animalium* (especially) that discuss the differences in the 'manner of life', 'activities', 'character' or 'dispositions' (*ēthē*) of animals, their friendships and enmities, whether they are gregarious or solitary and so on.[3] There, in the context of his zoology, we often find Aristotle speaking of the *intelligence* of animals (*phronēsis, phronimoi*)[4] and indeed using terms that might imply *moral* character – except that when he speaks of the 'courage' of the lion, that is, as the *Nicomachean Ethics* points out, only in an extended sense, since, strictly speaking, courage and the other moral excellences depend on *proairesis*, moral choice, and so cannot be ascribed to animals or to children.[5] Similarly when he speaks of some gregarious species as social, *politika* (e.g. *HA* 488a3ff., a9ff. mentions bees, wasps, ants and cranes as well as humans), that term does not imply that they form political associations in the strict sense that applies to the *polis*. Rather *politika* has the general sense in which it applies to any creatures, humans included, who act together as a group, engaging, as he puts it at *HA* 488a7ff., in one common (*koinon*) activity.

Aristotle is well aware that the relations between males and females vary very considerably across the animal kingdom (I shall be coming back to that again later). But that does not inhibit him from asserting that there is a natural association between the sexes for reproductive purposes, and he does not just mean natural *for particular* species. The key move that he repeatedly makes in this context is to identify the ideal or the goal as what is natural. So far

---

[2] One obvious example is the Great Speech of Protagoras in Plato, *Protagoras* 320cff. This is evidence of a pre-Aristotelian use of the *topos*, irrespective of the extent to which it represents the views of the historical Protagoras.
[3] e.g. *HA* 487a11ff., 488b12ff., 588a17ff. and extensively throughout *HA* ix.
[4] e.g. *PA* 648a5ff.: cf. Labarrière 1990.
[5] e.g. *HA* 488b16f. and the other examples at *HA* 488b13ff., with which compare e.g. *EN* 1116b24f., 30ff., 1118a23ff. cf. Fortenbaugh 1971.

as the animal kingdom goes, it is humans who provide the model to which other species approximate. Nature is a goal, *telos*, not just within a particular species (the perfect, complete, mature specimen contrasting with, on the one hand, the young and immature, and, on the other, with those who have passed their prime) but also *across* species, when human beings provide the standard by which other animals are to be judged.[6]

A desire to reproduce can, no doubt, be said to apply to all natural species, even while the ways in which that end is achieved vary: but that does not get Aristotle very far. In particular not to his conclusion that the most perfect, complete (and natural) mode of human social association is the *polis* – rather than any smaller or larger unit. Here it is of course Aristotle's view of the good life, and of the part that political and moral activity plays in that life, that provides the key. He states the basis of the argument for rejecting smaller communities, such as the village, in *Politics* book VII (1326b7ff.): the lower limit for the *polis* is set by the minimum number (of citizens) needed for self-sufficiency for the good life (crucially dependent, of course, on the leisure needed to engage in the cultivation of the moral, political and intellectual excellences of that life). But the upper limit to the size of the state is set, famously, by his notions of the directness of political participation: who could be general for an excessive multitude, who herald except some Stentor (1326b5ff.)? For good judgement in the matter of law-suits and in elections to magistracies the citizens must know one another: besides in vast communities foreigners and metics will infiltrate the citizen body (a remark that is all the more intriguing if it was made when Aristotle was living as a metic at Athens) and the impracticalities of the vast territory such a state would need are urged in objection to Plato's ideal in the *Republic* (1265a10ff.).

There are, then, specific political arguments adduced to support his conclusions concerning the upper and lower limits to the size of the *polis*. But that size, once determined, is not just what is right and proper, but what is natural. That is asserted, in book I, on the basis of his quasi-evolutionary account of the development of the *polis*: and again in book VII he bolsters his conclusions with

---

[6] For details of how humans are used as a model for other animals in the zoology, I may refer to my 1983, pp. 26–43.

an appeal to the general principle that there is a measure, *metron*, for the size of 'everything', to wit of animals, plants, and tools, *organa*: we may understand the products of nature and of art, but it is striking that the parallelism between the two is insisted upon. Excessive smallness or excessive largeness leads to either the object entirely losing its nature, *phusis*, or being inferior. A ship a hand's breadth long, or again one two stades in length, will be no ship at all, and even with recognisable ships, if they are too big or too small they sail badly (1326a35ff.). But while he does not say, in so many words, that there is a *natural, phusei*, size for a ship, he is prepared to use that locution of the state: how many and of what kind must the citizen body be *phusei* (1326a7)? Though *nomoi*, laws and customs, are of the essence of social and political institutions, those institutions are, in a way, just so many variants or diverse manifestations of the common, indeed universal, impulse, *hormē*, that humans have to form associations.

As so often elsewhere, therefore, there is a tension between *phusis* as goal, even as ideal, and *phusis* as what holds 'always or for the most part'. Aristotle knows very well that there are many other forms of human association besides the city-state: *it* is not *phusei* in the sense of what is universally, or even generally, the case, but rather in the sense, as he says, of the *telos*. On the other hand that is not *all* there is to his talk of *phusis* in this context: not *all* of his points relate directly or indirectly to his more or less predetermined ideas about the *ideal* society. At bottom he would want to claim that some of the features he speaks of as natural do indeed relate to humans as the kind of creature that we are – not just that biological desire to reproduce (1252a28ff.) but also the impulse to form associations, to constitute indeed a moral community (1253a29ff.). To be incapable of forming such associations is to be scarcely human, like a beast or a god, as he puts it (1253a27ff.). At this point, at least, he would have some modern thinkers join forces with him, those, that is to say, who have insisted that there is a minimum size to the social unit that can survive, for example to the primitive horde of hunter-gatherers – however far *they* may be from Aristotle's conception of the good life. Though individuals who have been brought up in society can survive on their own, Robinson-Crusoe-like, maybe indefinitely, there is nevertheless in the final analysis a truth in the dictum that the human being is a *politikon zoon* in the sense of a *social* animal, even if not exactly in Aristotle's sense of a *polis*-forming animal.

The second example I originally mentioned is what I called the notorious instance of the naturalness of the institution of slavery, by now the subject of a vast secondary literature.[7] Aristotle is well aware of a contrary view: for some, he says (1253b20ff.), mastery is contrary to nature, *para phusin*, for it is by *nomos* that one is slave, another free, and *phusei* there is no distinction, and so they believe that it is not just either, for it is based on force. Evidently when the naturalness of what Aristotle claims to be natural had been flatly denied, Aristotle has that much more of an obligation to *show* its naturalness, and he concedes straight away (1255a4ff., b4ff.) that there is a sense in which his opponents are right – though that is just because 'slavery' and 'slave' are ambiguous, being used of those deemed slaves *nomōi*, in contrast to the sense Aristotle wants, where some are slaves 'everywhere' (1255a31ff.). He can and does concede that some have become slaves unjustly, but that leads to no concessions on the principle that the distinction between slave and master is a natural one.

For his own, clearly contested, position the arguments he adduces (such as they are) consist first of a battery of proportional analogies. Master to slave is analogous to soul to body, to old to young, and especially to male to female (1254b4ff., 13ff., cf. 1332b35ff.). Yet, as so often, the argument is not merely one by analogy, for all these paired relationships exemplify and are instances of the generic distinction to be found, he says (1254a28ff.), in every composite whole constituted of parts whether continuous or discrete, namely that between what rules and what is ruled.

Once again the determination of 'the natural' involves a biological and a sociobiological dimension, but the way the evidence, or at least 'what happens' (*ta ginomena*), is put to work is striking. He can, of course, rightly claim that some distinction between soul and body is to be found in all living creatures: there are no problems there. But as for his representation of the relationship between males and females, Aristotle the zoologist knows very well that there are plenty of species of animals where the females are stronger, bigger, dominant, longer-lived than the males, and

---

[7] See Schofield 1990 and Garnsey forthcoming.

even one or two (bear and panther[8]) where they are more courageous. If we add up the exceptions that Aristotle allows to the rule that males are larger and longer-lived than females, these amount to a considerable list, one that includes most of the ovipara and larvipara according to *HA* 538a25ff. Yet that does not stop him from reiterating that naturally, *phusei*, and so to speak for the most part, males are longer-lived than females (*Long.* 466b14ff.). We may compare the even more extreme cases of right-handedness and uprightness, where Aristotle asserts that in humans alone the upper part is directed towards the upper part of the universe, and in humans the right side is most right-sided, and where he does not hesitate to propound the seeming paradox that in humans *alone* the natural parts are according to nature, *ta phusei moria kata phusin echei* (*PA* 656a10ff.).[9] What is normal or natural in the animal kingdom is evidently no matter of counting heads (or species) but is arrived at by reflection on the principles instantiated in the animals assumed to be *superior*, more particularly in the species assumed to be supreme, humans providing a model for the rest of the animal kingdom, to which however they only approximate – and all fall short to a greater or lesser degree.

It is a similar set of considerations that provides the key to Aristotle's pronouncements on slavery, for here what he claims to be natural for humans is a feature for which there is no parallel in the rest of the animal kingdom. A few species are said to have leaders, *hēgemones*, e.g. at *HA* 488a10ff. (though other gregarious animals are 'anarchic') and in the quite exceptional case of bees he is happy enough to speak of their rulers (as kings, of course, *basileis*, not as queens, e.g. *GA* 759a20). Yet for other species *not* to differentiate between ruler and ruled in this *extra* way in which humans do (in the master-slave relationship) is a mark of their inferiority. We should recall what he has to say about sex differentiation. It is *better* that male and female should be separated. At *GA* II 1, for instance, having said that male and female are formed out of necessity, Aristotle insists further that it is *better* that they should be separate. Why? Because that which possesses the *logos* and the form, namely the moving cause, is *better* than the matter, and it is *better* that the superior should be separated from the

---

[8] e.g. *HA* 608a33ff.
[9] cf e.g. *HA* 494a26ff., 33ff., *IA* 706a16ff., 20ff., b9f.

inferior. By parity of reasoning, therefore, he could claim that the slave/master distinction is a mark of human superiority (though I know of no passage where that is spelled out explicitly). Clearly for the superior to be separated from the inferior allows the superior to fulfil its potential: that that fulfilment is bought at the price of the incapacity of the inferior to fulfil any more than its, inferior, potentialities does not reduce the net advantage to zero (not that Aristotle thinks in simplistic quantitative terms in this regard). Rather he deems the advantages clearly to outweigh the disadvantages.

At this stage it may look as if Aristotle not only plays fast and loose with such sociobiological and biological data as he cites, but also is prepared to describe as 'natural' in the political domain more or less whatever happens to be the case in the particular type of society he happened to live in. Yet of course that is not true – as my third example underlines, an example that also highlights the problems he encountered at the borderline between the natural and the non-natural. This was Aristotle's attempt, in the latter part of book 1 of the *Politics* especially, to treat some part of the art of household management, *oikonomia*, *oikonomikē*, as natural. *Politics* 1256a1ff. raises the question of property and of wealth-getting (*chrēmatistikē*) and he is initially confident that the latter is not to be identified with *oikonomikē*, since it provides, while *oikonomikē* uses, goods. At 1256a19ff. we have, once again, a sociobiological argument: there are many different sorts of food or nourishment, and the ways of life of different species of animals (including humans) are differentiated according to their ways of getting food. It is nature, *phusis*, indeed that has so distinguished them (1256a26f.). That provides an over-arching principle, but then there is an analogy. Just as differences in what each kind of animal finds pleasant are according to nature, *kata phusin* (he instances carnivores and herbivores) and their ways of life vary accordingly, *so* the ways of life of humans too differ. That makes humans special – not that we alone are omnivorous: Aristotle knows that many crustacea, birds, snakes and insects, not to mention vivipara, are.[10]

Most of the human race, we are told (1256a38ff.), live from the land and from the fruits of cultivation. But the other ways of life he specifies as – it seems – *natural* ways of obtaining food con-

[10] e.g. *HA* 590b10, 594a4f., 596b10f.

tain some surprises. There are herdsmen (no problems there) and hunters, but the hunters include not just fishermen and hunters of birds and wild animals, but also those who live from brigandage (*lēisteia*, 1256a36, *lēistrikos*, 1256b5). There then follows (1256b10ff.) the famous argument to the conclusion that the rest of nature is in a sense for *our* sake. Again it starts sociobiologically: nature provides for the young of each species of animal in different ways, insuring that enough nourishment is available, in the larvae and eggs in the case of larvipara and ovipara, and in the form of milk for vivipara. So (he continues) nature also must be thought to provide also for the animal when grown up, and what that means is that plants are for the sake of animals (viz. presumably the herbivores) and the other animals for the sake of humans, both the tame animals, for service and food, and the wild ones, or most of them, for clothing, tools and suchlike.

The eventual conclusion that Aristotle reaches at 1256b26ff. is that there is one kind of art of acquisition, *ktētikē*, that is according to nature a part of *oikonomikē*, and I take it that that 'according to nature' is not just a matter of it being *appropriately considered* to be such. I take it, in fact, that we should take *kata phusin* strictly as in accordance with what is natural, since he has just argued that even warfare is in a way, *pōs*, a natural, *phusei*, art of acquisition (1256b23).[11] Starting from the innocent-looking observation that the acquisition of food is natural – indeed secured by nature for each kind of creature (1256b7ff.) – he eventually allows in quite a variety of modes of human acquisitiveness. These include hunting, but not just for animals, for food, but for other property as well: what he has principally in mind becomes clear in the sequel. When warfare is directed to subjugating those humans who are designed by nature to be ruled, *pephukotes archesthai*, then that kind of war is naturally just, *phusei dikaion* (1256b23–6). We should, to be sure, bear in mind what is said elsewhere about war, especially in book VII, where he insists that war is for the sake of peace (1333a35) and that the proper end of military training is not to enslave those who do not deserve it, but rather to avoid being enslaved yourself (1333b38). There too, however, it is recognised

---

[11] Contrast, however, Newman, 1887–1902. Yet at *Pol.* 1257a3ff., in the discussion of the kind of *ktētikē* that is more properly called *chrēmatistikē*, Aristotle distinguishes between a branch that is *phusei* and one that is not.

to be a legitimate end of warfare to rule those who *do* deserve to be slaves: 1334a2 makes explicit the point left implicit at 1333b38 (and cf. also 1255b37ff., where he refers to the *just* art of acquiring slaves).

The modes of acquisition that are described as natural thus range far beyond the mere provision of food. The state, he says, must either have or supply not just necessary goods, but also those useful for the community (state or household, 1256b27ff).[12] But evidently that is a very wide door indeed to open, and the question immediately arises of how Aristotle can conceivably *deny* the claims of a whole lot of other arts of acquisition to be legitimate, even natural. That he is himself preoccupied with just that question becomes clear immediately in the text: for after he had claimed various aspects of *ktētikē* to be natural, he immediately sets about contrasting those with *chrēmatistikē*, wealth-getting. One of these two is natural, the other not, 1257a3ff., but carried on by a certain skill or art.

The argument is a tricky one and seems to involve a shift or modification in his view of *chrēmatistikē*. He contrasts using shoes as shoes with using them as articles of exchange. Not that all of the art of exchange (*metablētikē*) is non-natural. No: even here there is a category of exchange that is not contrary to nature, but 'directed to the replenishment of natural self-sufficiency' (1257a30). It is natural that some people have more of some things, others of others. Exchange of goods for goods between one community and another, for that purpose, is *not* part of *chrēmatistikē* (1257a29): at least it is to be contrasted with the getting of wealth for its own sake, using coinage as the medium of exchange. Yet in the outcome (1257b19ff., 30ff.) he suggests a subdivision *within chrēmatistikē*:[13] 1258a5f. speaks of *heteron eidos tēs chrēmatistikēs*, and 1258a14ff. contrasts the necessary and the non-necessary, where the necessary kind is redeemed after all as naturally a part of household management.

---

[12] The text and its interpretation: see most recently Natali 1990. The disputed points do not, however, affect the substantial thesis that Aristotle proposed concerning the need for these goods.

[13] It is true that *Pol.* 1257b19ff. is potentially ambiguous and that the text of 1257b30ff. is in doubt (where some editors retain, but others omit, a negative before *chrēmatistikēs*, see Newman). However, the subdivision within *chrēmatistikē* is clearly referred to at 1258a5f., 14ff., and in the outcome a third type is introduced, as such, at 1258b27ff.: see below p. 194.

There is no weakening in his position on the status of wealth-getting for its own sake: that is condemned as having no natural *peras*, limit. There is no suggestion of compromise on *kapēlikē*, described as a kind of exchange that is justly censured (1258b1: since it is not in accordance with nature, but merely involves people gaining from each other). Yet although he had begun by distinguishing two kinds of *ktētikē*, contrasting one that is part of *oikonomikē* with *chrēmatistikē*, 1256b26ff., b40ff., by the end we have been given a more complicated analysis in which one kind of *chrēmatistikē* is contrasted with another, where one of those two kinds is considered *part of oikonomikē*, 1258a27ff., 38ff., even while the other (*kapēlikē* included) is still excluded from it.[14]

Taking stock of this convoluted discussion we may say that as with the first two examples I took, of humans as naturally social animals, and of slavery as a natural institution, the basis of some of the talk of the natural parts of household management is provided by what I called the sociobiological argument. The political analysis *starts* with what is true of humans as animals, the need for food. But if it starts there, it certainly does not end there. Aristotle allows the provision of other necessary goods and property – including particularly slaves (provided always that they deserve to be such): he even allows that the state has to have a provision not just of necessary but also of useful items. That, he claims, 1256b30f., is where true wealth comes from. But once those concessions have been made, he has some difficulty in *drawing the line* between the true and natural art of wealth-getting as he construes it and wealth-getting as it was normally understood by his contemporaries, to which there was, as he puts it (1258a18), no limit.

After all, if he allows warfare in order to enslave those who deserve it, then warfare to acquire agricultural land or land with particular natural resources, such as minerals, could (one might have thought) be justified equally well, or equally badly. It turns out that wealth-getting from mining is an intermediate, third, class between the type that belongs to household management and the type that is unlimited (1258b27ff.). But the line is not drawn, as one might perhaps have expected, between acquisition of *food* and *all* other modes of acquisition. Certain types of exchange, of goods for goods, are according to nature. Moreover from that beginning

[14] Cf. Natali 1990.

the employment of money, coinage, came to be devised out of *necessity, ex anankēs*, 1257a33f. But instead of allowing something to the argument that the increased facility provided by coinage for economic exchanges has a contribution to make to the good life (even as he, Aristotle, construes it: that will after all include the practice of such moral excellences as generosity), he focuses rather on the threat of wealth being accumulated indefinitely for its own sake.

The texts I have reviewed provide three of the main explicit appeals to the category of the natural in the *Politics*, but that of course is far from exhausting the occasions when ideas derived from his reflections on the world of natural phenomena are influential in his discussion of political questions. To do justice to that theme would require a lengthier discussion than space allows here, a discussion that would have to take into account, among other things, his frequent references to the characters that different citizens do or need to display as their *natures, phuseis*. But one particularly striking and indeed baffling feature cannot be passed over. I refer to a variety of ways in which political constitutions are treated as if they were – or could be compared with – natural kinds. First there is the extraordinary comparison between the classification of constitutions and the classification of animals in *Politics* IV 1290b21ff. There Aristotle proposes that the various types of political constitutions can be arrived at by identifying, first, the essential parts of a constitution[15] (on the analogy of the essential parts of animals, such as sense-organs, organs of locomotion and so on) and then determining how many combinations they form. This passage has always struck me as having important, if surprising, implications for Aristotle's thinking, at the time, on the topic of zoological classification – for his explicit aim here is, as he says, to grasp the kinds of *animal*, not just to give a check list of the differentiae or of the different parts themselves. However even the idea that he has an interest in zoological taxonomy inside, if not outside, the zoological treatises, is now highly controversial. For now however I do not need to rehearse the points at issue in that much contested debate.[16] For present purposes the important point is this: the very possibility of developing such an analogy depends

---

[15] Cf. also *Pol.* 1328b24ff.
[16] See especially Pellegrin 1982, 1986 and 1990, Lennox 1987b, Balme 1962a and 1987a.

on accepting two assumptions, first that the essential parts of a political constitution could be identified, and second that political constitutions themselves form, as it were, natural kinds.

To be sure in practice Aristotle's thinking about the different types of democracy, oligarchy and so on is complex, flexible and sophisticated: nevertheless his appeal to the zoological analogy would have no foundation unless his general assumption were that in politics too something like species are there to be found. Moreover the same assumption can be followed up in another feature of his analysis of political constitutions, namely his contrast between *normal* and *deviant* constitutional types, the three normal ones being monarchy, aristocracy and *politeia* (constitutional government) itself, the three deviant varieties, or *parekbaseis*, tyranny, democracy and oligarchy. The *terms parekbaseis, parekbainō*, themselves do not suggest a zoological model: the underlying concrete image is that of stepping aside, deviating, digressing, transgressing. Yet they are clearly said to be contrary to nature, *para phusin*, for example at 1287b39ff. The norm they deviate *from* is clearly conceived as *kata phusin*, and specific comparisons, for example with the different degrees of deviation of a nose (1309b23ff.) reinforce that general impression.

Finally the notion of *natural* norms also underlies the very frequent appeal to the metaphors of political and moral *sickness*. True, talk of the health of the state, and of moral health, and of the need for cures for the individual's or the state's moral or political ailments, had been common enough in Greek moral and political theorising from the beginning and had formed a major theme of Plato's contributions to that subject especially. But that does not mean that we should underestimate the implications in Aristotle's case, not just in his ethical treatises but also in the *Politics*. The notion of the *morbidity* of the constitution may be close to being a dead metaphor, as also may talk of remedies or cures in the political context. But we find also some full-scale analogies developed on their basis – for example at 1320b33ff. on the topic of the political constitutions that are particularly liable to weakness and that therefore need special care to maintain (and cf. 1284b15ff., a text that uses a distinction between a state that is self-maintaining, since it has a healthy constitution, and one that requires the *iatreia* of the law-giver). At least we have to bear in mind that such metaphors, images and analogies secure whatever plausibility they may have from the ease with which the transfer of ideas from natural organisms to political constitutions may occur.

At this stage we may pause and ask what we have learned so far from our inquiry into Aristotle's use of the category of the natural in the *Politics*. Two negative conclusions provide a starting-point. On the one hand, the use of that category is *not limited* to what is true (or what Aristotle believes to be true) just of humans *as animals*: it is not limited to points that apply also to other animals, for example those that are social. For certainly when slavery is held to be a natural institution, that does not apply to *any* other creatures besides humans.

On the other hand, the use of that category does not simply comprise whatever Aristotle believes to be the case with *all* human societies, *or* whatever he happens to approve of in that regard. So far as the former point goes, the deviant constitutions make the point clearly enough: there are forms of human association to which Aristotle would certainly not be prepared to attach the label 'natural'. Again plenty of human societies exemplify non-natural or unnatural economic activities (such as retail trade) of which the same is true, namely that Aristotle would refuse to describe them as natural. But as far as the second point goes, it is important to observe that (although the two often coincide) Aristotle does not straightforwardly equate the human political and social institutions he approves with what he is prepared to describe as natural. Much of the description of the ideal state, for example of its educational system, goes beyond anything for which he claims or could claim the property of naturalness, and of course at one level there is a distinct contrast, in the *Politics* as in the ethical works, between, on the one hand, *phusis* in the sense of the natures or characters of individuals, and, on the other, training or instruction (*didachē*) and the effects of habituation (*ethos*) (with *EN* 1179b20f. compare, for example, *Pol.* 1332a39ff.).

For the label 'natural' (*phusei, kata phusin*) to be applicable to political institutions or behaviour it would seem that – often if not always – two criteria have to be fulfilled. First the items in question will be ones of which Aristotle approves (that is, not just what he recognises to be common): thus unlimited wealth-getting may be recognised to be widespread, but it is not natural, indeed it is contrary to nature (as we saw) *as* unlimited.

But secondly, in the three main cases we began with, at least, the institutions or behaviour should exemplify, or be grounded in, some general principle that can be held to apply elsewhere than just to political life, viz. for instance in the animal kingdom, if not more generally still. Mastery-slavery is (we said) only exemplified by humans, not by other animals: but it can be seen as a mode of ruling and being ruled and *that* principle *is* natural in the sense of applicable also to a wide range of other contexts besides politics, being exemplified in different ways also by the relationships between soul and body, old and young, male and female. Those other cases we considered, where the notion of natural kinds influences his thoughts on the classification of political constitutions, are rather more complicated but may also be thought to fit the same schema. Certainly the perverted constitutions fail the test of the natural in that they are clearly disapproved of: the constitutions that are according to nature, conversely, are those that provide instances of possible, correct, therefore natural, modalities of ruling and being ruled.

Be that as it may, it is clear enough and would not be disputed that throughout their use in the *Politics* there is a tension between the descriptive and the normative uses of nature and the natural, and this takes me now to the second of the three main topics I identified at the outset. It is true that, for someone who comes to the *Politics* on its own, the normative role of appeals to what is natural may come as a surprise and even be rather shocking. It is not just that Aristotle seems to fail to observe a due distinction between the realms of nature and of human society, laws and conventions – between *phusis* and *nomos* – but he believes, it seems, that he can justify some recommendations with respect to the latter by appeals to the former. But as any student of Aristotle knows and as I have already had occasion to mention, it is an all-pervasive feature of his thought that *phusis* is *both* descriptive *and* normative. There is nothing covert about this. On the one hand what is natural corresponds to what is regular, what is usually the case – as he repeatedly expresses it, what holds 'always or for the most part'. On the other, nature is clearly, repeatedly and explicitly a *telos*, end, goal, the *hou heneka*, that for the sake of which (in the sense of end or goal). That is spelt out *expressis verbis* for example at *Physics* 194a28ff. and again at *Physics* 199a30ff.

But that those two features of his use of *phusis* can readily be exemplified throughout the Aristotelian Corpus says something, merely, for his *consistency*: it does nothing to justify that use or defend it against possible charges of confusion. Let me begin with one difficulty that can be identified most clearly in his natural philosophy. Evidently 'nature' is often invoked to discount certain parts of the phenomena, where Aristotle focuses on some part of what is, in some sense, given as what is important while dismissing other features as unimportant ('noise'). The difficulty arises in that, depending on which of the two features of *phusis* is in play (or dominant), the principles for the discounting will differ. Thus if he finds an anomalous individual spider or octopus or whatever, anomalous, that is, in its anatomy, morphology, physiology or even in its ways of behaviour, he rightly discounts it to focus attention on the properties of the overwhelming majority of members of the species in question: that presupposes, to be sure, that the species can be identified as such and that may prove more difficult than Aristotle sometimes assumes. Yet in many cases there will be agreement enough about what counts as the natural kind to allow the distinction between the ordinary and the anomalous specimen to be drawn at least as a first approximation. As he puts it in the *Politics* itself, 1254a36ff. in dealing with what is according to nature one should consider what is *phusei* rather than the *diephtharmena*.

On the other hand when 'nature' is equated with the goal, he can and does discount far more than the exceptional features of particular anomalous specimens. The whole of the female sex – to revert to that notorious example – is as it were a 'natural deformity' (*anapēria phusikē*).[17] In that expression 'natural' corresponds to the notion of nature as what is regular (always or for the most part), for there is nothing unnatural about the females of the species in that sense of 'natural'. But 'deformity' just as surely invokes the idea of nature as *telos*. Females are in that sense deformed, in that they fail to concoct the blood (in the way males do), or, otherwise said, they fail to be males.

The difficulties can be identified, and, to be sure, not all of them can be surmounted. Yet *some* defence of *some* general aspects of Aristotle's use of the concept of *phusis* can be offered. To begin with those cases of discounting in zoology that I have just mentioned. There is nothing objectionable in the idea of treating zoological anomalies as just that, as anomalies. No doubt we could

---

[17] e.g. *GA* 775a15f., cf. *pepērōmenon*, *GA* 737a27ff., cf. above ch. 4. at pp. 91f.

invoke a more sophisticated justification for distinguishing between the normal and the abnormal. Where appropriate we can and should quantify degrees of probability and of course we have a rigorous theory of error. In practice such a standard textbook as the *Biology Date Book* of Altman and Dittmer normally and for preference gives upper and lower limits for 95 per cent of the given population for the physiological quantities it specifies, at least when dealing with large populations.[18] No doubt the notion of a species is itself problematic in ways that Aristotle would never have imagined: and no doubt what is true of members of whatever species is identified would be reported in altogether more rigorous ways. But it remains true that there is a legitimate, even essential, role for the distinction between what is true 'for the most part' and anomalies or exceptions.

Of course when we turn to Aristotle's comparisons *across* species and his appeal to higher creatures, humans especially, as the norm to which other, lower animals only approximate, many of his specific examples appear to be misguided. Certainly, too, there is as much or as little to be said for considering males as deformed females as for treating females in the way Aristotle does. Yet when many concessions have been made, it is not as if the principle of cross-species comparison is *in itself* mistaken or inevitably misleading. When Aristotle calls the mole deformed (*pepērōmenon*) because he believes its residual eyes do not function as eyes, what is wrong is not the attempt at the comparison, but the biological facts of the case.[19] His comparisons between the arms of humans and the forefeet of quadrupeds, and again between the latter and the wings of birds,[20] are not *in principle* objectionable, though we may quarrel with many of his particular interpretations, and we now have, and would want to use, a clear distinction in this context between homologies and analogies.

But may it not still be argued that Aristotle plays fast and loose with the contrast between 'is' and 'ought' statements and even with the traditional Greek contrast between *phusis* and *nomos*? To take the second of these two points first. It is clearly not the case that Aristotle abolishes that contrast. That some are slaves *nomōi*

---

[18] See Altman and Dittmer, 1972–4, vol. 1, pp. xvf.
[19] See *HA* 533a2ff., cf. *de An.* 425a9ff.
[20] e.g. *PA* 693a26ff., *IA* 712b22ff.

who are not *phusei* reminds us of that. Yet he has certainly extended the notion of what is true *phusei* so that it covers some institutions or behaviour that were normally considered to be matters of convention or custom – in some cases *only* or *merely* matters of convention or custom. But in some such cases the area Aristotle deems to be encompassed by *phusis* includes some items that are in his *own* view *also nomōi*: in other words those two terms are not mutually exclusive, but rather there is an area of overlap between them. Again the example of slaves and some of the economic analysis we considered can be used to make the point. For when natural slaves are *also* slaves by law (*nomōi* in that sense), *phusis* and *nomos* coincide: and they also coincide when *natural* exchange or barter is *nomōi*, i.e. customary. The *nomos/phusis* contrast had, in any case, been construed in quite a number of different ways by different Greek thinkers. Plato in particular had already made moves to subsume some parts of *nomos* under *phusis*, notably in *Laws* x where he resists those who straightforwardly opposed those two and insists that human *nomoi* and *technai* are or should be the products of *nous*, reason, and therefore of a principle analogous to that on which the universe as a whole depends.[21] In this context Aristotle too may be represented as redefining the *nomos/phusis* contrast in such a way as to allow some of the former to be subsumed under the latter, though his grounds are different. It is not that *nomos* can evince the same demiurgic intelligence that is at work in the cosmos as a whole: rather that some (but of course not all) human customs, conventions and laws can be said to be according to nature, exemplifying, in particular, *natural* modalities of the ruling/ruled relationship.[22]

And so finally I return to the is/ought contrast and to the last of the major questions I identified at the outset, the evaluation of Aristotle's position on the fact/value dichotomy. Until fairly recently much of the English-speaking philosophical tradition treated Hume's principle as fundamental, if not sacrosanct: but the situation today must be recognised to be more complex. First there is the relatively minor technical point that Searle and others have investigated the conditions under which it *is* permissible to derive

---

[21] See Plato, *Laws* 890d, cf. 889aff., 892b, 896eff.
[22] Art in general imitates nature, as Aristotle several times proclaims, *Phys.* 194a21f., 199a15ff., *Mete.* 381b6, and of course analogies between the two domains are a recurrent motif (some are documented in my 1966, pp. 285ff.).

'ought' statements from 'is' ones:[23] however the conditions are generally such that the point remains largely a technical one.

But then secondly and more importantly, the fact/value dichotomy has been eroded at least in certain contexts, notably by points developed in recent philosophy of science. Observation statements, and factual statements more generally, always presuppose a conceptual framework, and any such framework implies choice, selectivity, a view-point. There cannot then, so it is argued, be any discourse that is totally value-neutral, totally value-free. In particular science, as inevitably theory-laden, is inevitably evaluative, at least in some loose sense of evaluative. This is not to deny that the observation/theory and the fact/value contrasts are needed to convey important and valid distinctions: on the contrary we can clearly not dispense with them. But those contrasts should rather be construed as matters of *degree*, not as contrasts between mutually exclusive terms. Not only do theoretical statements and observation statements not form two mutually exclusive sets: but *pure* observation statements (if not also pure theoretical ones) are not to be found. Rather the contrast that is valid and important is one between *greater* and *less* theory-ladenness, and again between greater and less *value*-ladenness.[24]

But if many of us nowadays would insist that the old-fashioned positivist assumptions about a value-free, theory-free natural science have to be abandoned, where does that leave the evaluation of Aristotle's position on the relationship between natural science and politics? What from that old-fashioned point of view seemed so scandalous, namely Aristotle's dangerous confusion of the descriptive and the normative, his readiness to treat sociology and zoology as *similar* studies, his use, precisely, of a category of nature and of the natural that *spans* politics (or parts of it) *and* natural science, can now be reassessed.

Three points stand out. First it is as well to remind ourselves that Aristotle still *has* distinctions, and very firm ones, between moral philosophy and natural science. *Natural* excellence may span the animal kingdom as well as humans: but moral excellence, depending on *proairesis*, is limited to humans.[25]

---

[23] Cf. Searle 1964.
[24] There are references to recent philosophical discussions in my 1979, p. 128, and 1991, ch. 15.
[25] The classic text for *phusikē aretē* is *EN* 1144b1ff.

Where, however, secondly, Aristotle does want to provide a foundation *in nature* for some of his analytic and prescriptive remarks about human social and political institutions, we must still express our reservations. If some of his biological and sociobiological arguments have *some* force, the hierarchisation that he discovers permeating nature depends crucially on evidence that is not just highly selective but heavily interpreted. Thus what might have been deemed counterevidence from his zoological researches to his view of the relationship between the sexes, for instance, is all too easily accommodated, by appealing to the further assumption of the inferiority of some species of animals to others, the existence of whole species that are 'deformed'. Moreover reflection on some latter-day attempts to draw moral implications either from biological theories (the selfish gene) or from sociobiological ones (the naked ape) confirms the dangers of such endeavours – the distortions involved if not within the scientific speculations themselves, certainly in their application to humans and to human societies.

Yet, thirdly, the reflection on those difficulties and on the parochial nature of many of Aristotle's suggestions is no grounds for complacency. It is of course much easier to criticise and condemn specific efforts to resolve the problems of foundational principles in ethics and in politics than it is to offer good-looking solutions to those problems oneself. Some would go so far as to say that the urge to seek foundations is the root of the difficulty, a point of view with which it is easy to sympathise if 'foundations' are construed as some kind of objective basis for a universal ethics (though with or without foundations, the question of justification remains).[26]

At the same time relativisation to a group, a community, a society as a whole, even, also has its attendant disadvantages. No doubt the styles of naturalistic ethics repeatedly proposed in antiquity, Aristotle's included, now appear at many points naive, not least in their understanding of human psychology. But the mistake is not so much thinking that there may be some relevance in science and cosmology to our understanding of how we should live, as in what was actually believed under the heading of science and

---

[26] Various aspects of problems to do with foundations in both ethics and philosophy of science have been discussed by Feyerabend 1961 and 1975, Hesse 1980 and Rorty 1980.

cosmology. Ethical recommendations are, for sure, not going to be read off or directly derived from the findings of natural science. Yet as I have argued elsewhere,[27] science and cosmology have things to teach us that are certainly relevant to how we should behave, that are relevant to our *self*-understanding as well as to our understanding of what is other, the world of natural phenomena. It is after all science and cosmology that have contributed to releasing us from some of the anthropocentric assumptions so deeply engrained among the ancients (not that we are by any manner of means free from such assumptions ourselves). What is needed, as always, is better science and cosmology, a better understanding of *phusis*, including the nature of man, but that does not mean (it cannot mean) a science devoid of relevance to questions of values. While we may often suspect Aristotle of a certain degree of opportunism in the conclusions he carried over from zoology to politics, there is more to his understanding of what those studies have in common than simplistic condemnations based on invocations of Hume would allow.

[27] Lloyd 1991, ch. 15.

CHAPTER 10

The metaphors of metaphora

Aristotle's attitudes towards what he calls *metaphora* (not exactly equivalent, by any means, as is well known, to what we mean by 'metaphor') exhibit what appear to be certain deep-seated ambivalences. On the one hand, he is critical, even at points contemptuous, of their use. On the other, there are passages where he appears to applaud and appreciate their deployment.

My aim, in this chapter, is to explore just one aspect of this ambivalence. It is a striking feature of his explicit account of *metaphora*, in the *Rhetoric* and *Poetics*, that it relies very heavily on the very language-use that Aristotle's official theory condemns. Why is there so much metaphor, so much *metaphora* indeed, in Aristotle's theory of *metaphora*? Some would say that that was inevitable, but that is not Aristotle's usual view. What did Aristotle himself think, for he surely did not fail to notice? What can we learn, here, about his own use of metaphor?

The suggestion I wish to explore raises a general question about his theory of literary criticism and starts from a dilemma. If Aristotle's theory of literary criticism is itself, broadly, a part of rhetoric, then that would help to explain his own heavy reliance on metaphor in his account of *metaphora*. Yet if we take that view, we would have to broaden, rather, the category of rhetoric, at least by comparison with some of his characterisations of it.

If, on the other hand, we deem his theory of literary criticism to be part of philosophy – that is of practical philosophy – or of dialectic at least, then the implications of his *use* of metaphor, in his account of *metaphora*, would be that we have at least an implicit approval of their use.

But on either hypothesis, Aristotle turns out to be a more positive practitioner of metaphor than his official theory of language seems to allow us to expect.

I shall proceed first by setting out some of the evidence both for the negative and critical attitude and for the positive and appreciative one. Some of these texts are extremely well known and I have commented on some of them in earlier studies: so I must ask indulgence for some parading of familiar points. However, only if that material and those points are before us can we hope to tackle the further specific questions I wish now to raise in relation to the use of metaphor in the account of *metaphora*, where I hope to advance beyond my earlier forays even if into admittedly speculative terrain.

First we must be clear on how Aristotle uses the term *metaphora* and where it differs from what we might call 'metaphor'.[1] Aristotle's own definition in the *Poetics* (1457b6ff.), *onomatos allotriou epiphora*, 'the application of a strange term', is a touch Delphic and very general, though it establishes that a *metapherein*, bearing across or transfer, is an *epipherein*, bearing to or application. The famous four-fold classification that follows – from genus to species, from species to genus, from species to species and by analogy in the sense of based on four-term proportional analogy, as A is to B, so C is to D – has a rather scholastic ring. But given that 'from species to species' is potentially a catch-all category, the classification will serve very generally to identify possible types of transfer.

Both the first two types indicate, however, that the transfers that Aristotle would include under *metaphora* are far wider than just those that involve a direct explicit or implicit comparison between two terms. Thus the second type, from species to genus, is the use of a specific term in place of a more general one. But the example he gives to illustrate this is 'indeed Odysseus did a myriad deeds', where 'myriad' is the 'specific' term applied in lieu of the more general 'many'. That evidently involves no *comparison* between what is here thought of as the *eidos* and the *genos*, even while it clearly meets the Aristotelian criteria of the 'transfer of a strange term – viz. of one that belongs to something else'.

There is a further important difference between some later analyses of metaphor, simile, comparison and what we find in Aristotle. Later analyses tend to take comparison as the generic term

---

[1] Among the classic discussions, those of Ricoeur 1977, and Derrida 1982, remain fundamental. Among recent work I have found the paper of Laks in the recent Symposium Aristotelicum collection on the *Rhetoric*, edd. Furley and Nehamas, Laks 1994, particularly helpful.

and see metaphor and simile as species of it – and as between those two often interpret metaphor in terms of simile rather than vice versa. In Aristotle, however, an *eikōn* is a kind of *metaphora*, as he tells us, so he says, on many occasions, *Rhetoric* 1406b20, 1412b32ff., 1413a13f. At *Rhetoric* 1410b17f., where the reading is in doubt, he says that the difference between them is a matter of a *prothesis* or a *prosthesis*. Those who read *prosthesis* have usually taken it that what he has in mind is specifically the addition of such a particle as *hōs*, as, like: earlier, at 1406b21ff. the difference had been illustrated as one between 'as a lion', *hōs de leōn*, and 'a lion', *leōn*, 'he rushed forward', said of Achilles.[2] Or else, if we read *prothesis*, the difference is a matter of either a 'setting forth' or a 'prefixing'. Be that as it may, the standard Aristotelian position subordinates *eikōn* to *metaphora* as species to genus.

However, it is clear that Aristotle's point of departure is not the use of comparison, likeness, resemblance as such, but rather a basic contrast between terms used strictly (*kuriōs*) or appropriately (*oikeiōs*) on the one hand, and terms *not so used* on the other. This presupposes that there *is* a domain within which the term *is* used strictly and appropriately and that is crucial, no doubt, to the model of language with which Aristotle works.

Within that domain, he does not appear concerned to distinguish between what we might call more, or less, central applications, though that is an idea in play in his analysis of 'focal meaning' (cf. ch. 7 at n. 1). But clearly anything that carries the appearance of not being a *central* application risks being ruled to be a *metaphora*. Nor is he concerned to analyse the phenomena of what we call dead metaphors, or the ways in which certain boundaries between what is and what is not perceived as a metaphor may change – even though he does show a reasonably acute sense of a number of changes in *style* that have resulted either from the development of poetic genres or from changes in fashion in rhetoric itself. But while he occasionally speaks of transferred terms as *more* and *less* strict, using the comparatives, *kuriōteron*, *oikeioteron*, as at *Rhetoric* 1405b11ff., for instance, it is not that strictness and appropriate-

---

[2] At *Rhetoric* 1407a13f. *eikones* are said to be *metaphorai logou deomenai*, where, if *logos* is 'word', the phrase *logou deomenai* has to be taken with *metaphorai* and read concessively (similes are metaphors, though the metaphors lack a word), if that text is to square with this reading of 1410b17f. At any rate, of the two, Aristotle regularly takes *metaphora* to be the more concise.

ness are always or even generally a matter of degree. Strictness, he would claim, can indeed be spoken of strictly: and what is appropriate need not be subject to any qualifications with respect to its appropriateness.

The model and the norm are provided by terms used *kuriōs*, and the *deviant* character of *metaphora* comes to be stressed very heavily in the context of his analysis of deductive reasoning and strict demonstration. What I called the negative attitude towards metaphora is prominent in the *Organon* and in certain contexts in the *Metaphysics*, though it is far from being confined to the logical and metaphysical treatises. Syllogistic, as presented in the *Prior Analytics*, presupposes the univocity of the terms used in universal or particular predications. Any suggestion of *metaphora* or of non-*kuriōs* usage when one thing is said of another would evidently undermine the validity of any chain of reasoning – and that will apply not just to syllogistic in the Aristotelian sense but to any deductive reasoning. In the account of the strictest style of demonstration, *apodeixis*, in the *Posterior Analytics*, the requirements on the premises from which demonstration proceeds include not just the univocity of the terms, but that the premises should be true, primary, immediate, better known than, prior to, and explanatory of, the conclusions – an ideal we have discussed in detail in chapter 1.

It is not at all surprising, therefore, that some of the warnings against *metaphora* are dire. In *APo.* 97b37ff., for instance, dealing with definitions, he remarks: 'if one should not argue in metaphors, it is clear that one should not use metaphors or metaphorical expressions in giving definitions'.[3] It is true that in the *Topics* and *de Sophisticis Elenchis* the use of likenesses (*homoiotētes*) is recommended for certain dialectical purposes. But one context in which that happens is when he is providing tips to achieve victory in debate. There are even occasions when he suggests how one can exploit likenesses to deceive an opponent. He does so, for instance, at *Topics* 156b10ff. and *SE* 174a37ff., and at *SE* 176b20ff. he says that it is particularly by using *metaphora* – by transferring, *metapherōn* – that one can make an argument difficult to refute. You

---

[3] The last phrase is sometimes taken as one of the objects of *horizesthai*, and the whole then translated: 'if one should not argue in metaphors, it is clear that one should not use metaphors in giving definitions, nor define metaphorical expressions'.

can avoid the appearance of sophistry since there is an unclarity about the truth of the matter: and you can avoid the appearance of lying since there is an ambivalence in the expression. The outcome is that *metaphora* will make the *logos anexelenktos*.

But although there are these tactical, dialectical uses of *metaphora* in the *Organon*,[4] *any* recourse to *metaphora* introduces an unclarity that is utterly inimical to the enterprise of strict demonstration – the drawing of true, incontrovertible, conclusions, by valid inference, from self-evident, indemonstrable primary premises. In the context of definition, again, in the *Topics*, he identifies the fundamental problem. One important point to study in examining a definition, he says, is whether it is said *kata metaphoran*: for every metaphorical expression is unclear (139b34f.).

Within the *Organon*, the negative comments made about *metaphora* would seem to stem from, and be explicable in terms of, the theory of the syllogism and of demonstration that Aristotle there puts forward. However as one pursues the topic in other texts in other treatises, where a similarly hostile viewpoint is adopted, other considerations also come into play, and in particular Aristotle's insistence on the boundaries between genres, and especially that marking the contrast between philosophy and dialectic on the one hand, poetry and rhetoric on the other.

In a variety of contexts in treatises dealing with very diverse subject-matter he criticises opponents or rivals, named or unnamed, for *their* use of comparisons, analogies, *metaphorai* and the like. This is not just a matter of his criticising certain comparisons as ill-formed, as superficial, based on deceptive similarities or on no similarities at all:[5] he also criticises them for the use of metaphor as such. That is the moral of the objection urged against Empedocles' remarking that the sea is the sweat of the earth. That may be adequate, Aristotle says, for the purposes of poetry – for *metaphora* is poetical – but it is not adequate for understanding the nature [of the thing] (*Mete.* 357a24ff.). Again, notoriously, when launching into an attack on Plato's theory of Forms (*Metaph.* 991a20ff., 1079b24ff.), Aristotle puts it that 'to say that they [the Forms] are models and that other things share in them is to speak nonsense and to use poetic metaphors'.

[4] Cf. below, at n. 15, on the use of likenesses in securing the genus of definitions.
[5] A series of examples is reviewed in Lloyd 1987, pp. 183ff.

The labelling of metaphor as 'poetic' has a two-fold significance for our inquiry. On the one hand, to depend on metaphor, so he would have us believe, in natural science and in metaphysics, is hopelessly inadequate. On the other, when we turn to the positive appreciation of metaphor, we had better be careful to see whether that approval is limited to the particular, generally inferior, genres in which it is most likely to be deployed, principally poetry itself.

Thus to follow up the first point, Aristotle is as keen as any ancient Greek writer to contrast the high style of philosophising that he advocates with the inferior would-be 'wisdoms' he associates with the poets and the rhetoricians, let alone with the 'sophists'. When a natural philosopher such as Empedocles chose to write in verse, he was especially vulnerable to Aristotle's criticisms. In that case, Aristotle has a field-day, scoring twice over. On the one hand, Empedocles is no good as a poet: he has nothing in common with Homer, as we are told at *Poetics* 1447b17ff., except the metre. On the other, he is no good as a natural philosopher either, not at least when he uses metaphors that are not adequate for understanding the nature of things. Again the criticism of Plato is particularly dastardly. Plato himself, after all, had distinguished his ambitions for philosophy, and what he calls the dialectical method, from the offerings of 'sophists', rhetoricians, and, precisely, poets, and he had, moreover, engaged in a devastating critique of more or less the entire legacy of Greek poetry. So it is particularly pointed that Aristotle should turn the tables and label one of Plato's central metaphysical notions (mere) 'poetic metaphor'.

But the second point of relevance of that collocation, 'poetic metaphor' takes us now to the positive evaluations that Aristotle sometimes expresses. Alongside what I have called the negative and critical attitude towards *metaphora* that Aristotle sometimes displays, there are signs of a more favourable view, at least in certain contexts and for certain purposes, especially in the *Rhetoric* and the *Poetics*. But the question we have to ask is: what contexts and what purposes?

One point that emerges clearly from such a passage as *Rhetoric* 1405a4ff (cf. 1404b33ff.) is that metaphor is recognised as an important resource both in poetry and in prose.[6] But as 1406b24ff.

---

[6] *logois*: but here and on other occasions in the *Rhetoric* and *Poetics* Aristotle is not concerned just with speeches, but also with written compositions as well (cf. below, n. 14) and so with prose in general as opposed to poetry.

goes on to observe about similes, while they are useful also in prose, they should be used less frequently – that is than in poetry. And the reason? There is something poetic about it.

It is particularly for the contribution that it can make to the excellence of *style*[7] that *metaphora* is valued. Skill in its use, he says in the *Poetics*, 1459a6ff., is indeed a mark of natural genius (*euphuia*), explaining that in turn with the remark that to use *metaphora* well is to observe likeness. At *Rhetoric* 1405a9f., too, he comments that to use *metaphora* well – to produce 'perspicuity, pleasure and a foreign air' – is not something that can be learnt from anyone else.

Thus far the positive and the negative attitudes towards *metaphora* might be reconciled with the following argument: the critical points Aristotle makes effectively exclude *metaphora* from serious philosophy – on the grounds of the unclarity it imports – while the positive evaluation is limited to its role as an ornament of style, especially though not exclusively in poetry. But if so, then the question as to why Aristotle should have used *metaphora* so extensively in his own account of it becomes more urgent and more puzzling.

We may now turn to our principal agenda, and we must first review some of the actual *metaphorai* Aristotle deploys in this context. We may begin, of course, with *metaphora* itself, since, as has often been observed, 'transfer' (the noun or the verb) is here transferred from its own strict application, which would be to the transport of physical objects, for use, in the 'strange' *allotrion* context of terms.

So too the term for 'strict', *kurios*, is used primarily of whoever or whatever has control or exercises authority. The substantive is often used of the head of the family, guardian or trustee, the lord or the lady of the house. The term for 'appropriate', *oikeios*, 'literally' means 'of the household', so is used of household property, or family or kin, and then of what is one's own more generally. Conversely the term *allotrion*, which I translated 'strange' in the *Poetics* definition of *metaphora* at 1457b6ff., means belonging to another, *allos*, and is often used of what is foreign.

The basic vocabulary for describing what *metaphora* is is thus full not just of what *we* might term the 'metaphorical' but of

---

[7] *lexis*, diction, speech, said by Aristotle in the *Poetics*, 1450b13ff., to cover the expression of meaning in words, is defined at *Rhetoric* 1403b15ff. to include the topic of *how* to say things in general.

what Aristotle himself treats as *metaphora*. The same is also true of his account of its effects and of how to use *metaphora* to the best advantage. In the text just quoted, *Rhetoric* 1405a8ff., where he says that the use of *metaphora* cannot be learnt from anyone else, he remarks that it produces 'perspicuity' (*to saphes*), pleasure (*to hēdu*) and 'a foreign air' (*to xenikon*), and both the question of perspicuity or clarity and that of foreignness call for some comment.

In both cases Aristotle seeks to identify an excellence of *metaphora*, though it is one that will involve a mean between extremes, and in both cases he relies heavily on what are, admittedly, very common metaphors. First, 'perspicuity' can be achieved by the use of *metaphora*, even though he had put it, in the *Topics*, that, from the point of view of definition, every metaphorical expression is obscure, *asaphes*, the precise antonym of the word translated 'perspicuity', *to saphes*. The positive quality is, no doubt, connected with another merit of *metaphora* that Aristotle speaks of, which we shall be discussing shortly, namely its ability to bring things 'before the eyes', that is, as we might say, its vividness. But for now it is enough to note that, for both the positive and the negative qualities, it is *vision* that provides the source of the characterisations. From one point of view, in philosophy, *metaphora* imports unclarity: but as a feature of literary style it can make things clear, though Aristotle goes on to note, *Rhetoric* 1412a9ff., that metaphors should be drawn from topics that are appropriate and *not evident* (*mē phanerōn*). Clearly, if clarity is a virtue, that does not prevent what is evident or obvious being a weakness.

The 'foreign' air, *to xenikon*, claimed as a merit of the use of *metaphora* at *Rhetoric* 1405a8ff. and elsewhere (e.g. *Poetics* 1458a22f.), may be thought to develop a feature already present in the use of *allotrion* (translated 'strange') in the definition at *Poetics* 1457b6ff. Once again, however, there is a tension. 'Foreignness' is a striking and worthwhile effect. Yet excess here too is a weakness. *Rhetoric* 1404b36 and 1405a8f. put the positive point, but 1406a14ff. warns, in the context of the use of 'epithets', that one should aim at the mean or the moderate, *to metrion*, for otherwise that leads to the fault of style that Aristotle calls 'frigidity' (*to psuchron*) – a particular danger in the use of *metaphora*, as 1406b4ff. elaborates.[8] While

---

[8] One may note the criticism of *metaphorai* that are 'too poetic' (*poiētikōs agan*) leading to 'frigidity' at *Rhetoric* 1406b10f. (cf. 1406a32 on 'epithets').

Rhetoric 1410b31ff. encourages the use of *metaphora* for effect, they should be neither strange (for then they are difficult to comprehend) nor superficial (for then they do not impress). But in this context it is said that one should avoid the strange, *allotrion*, despite the fact that metaphor is defined as the application, precisely, of a strange term.

The virtues of style in general are captured with an array of *metaphorai* in *Poetics* 22 and *Rhetoric* III 2ff., where differences between different genres are duly pointed out. Thus apart from needing to be 'clear' (*saphēs*), it should be 'not humble' (*mē tapeinos*), and both texts point to tensions there. First *Poetics* 1458a18ff. states that while 'strict' (*kurios*) terms are best for clarity, they conduce to 'the humble'. *Rhetoric* 1404b4ff. further points out that while the poetic style is 'not humble', it is inappropriate to prose. So again a balance needs to be struck, 'not humble', but 'seemly' (*prepousa*) or what 'fits' (*harmottei*), 'solemn' (*semnos*) and 'ornamented' (*kekosmēmenos*) as the genre and occasion demand.

Rhetoric III 10 and 11, discussing the advantageous effects of the use of metaphor, do so with a particularly striking use of metaphor themselves. The topic of chapter 10 is how to achieve *ta asteia* and *ta eudokimounta*. The former term is often translated 'witty' or 'smart'. But the primary meaning of the word from which the term is derived is 'city', *astu*, so these are the expressions one might associate with the city-dweller, before they come to be thought of as 'witticisms' in general. Again *ta eudokimounta*, translated 'popular' or 'well-esteemed', is a participle from a verb used of people of good repute.

But metaphors do not just contribute to 'city-talk' and to the 'well-esteemed': they are repeatedly said to be effective in bringing the matter talked about 'before the eyes' (*pro ommatōn*), *Rhetoric* 1410b33ff., 1411a26ff., 35ff., b4ff. At 1411b24ff. Aristotle feels the need to explain the expression himself, and he does so in terms of a series of further terms that relate to what is alive, as 'living' (*zōnta*, 1412a9), 'animate' (*empsucha*, 1412a2), or more generally 'moving' (*kinoumena* 1412a8) and most often 'active' (*energounta*, *energeia*, 1411b26, 28, 29, 30, 33, 1412a3, 4, 9).

Aristotle does not explicitly draw attention to the fact that his whole account of *metaphora* in general, and of its advantages and dangers in use, itself draws heavily on *metaphorai*. But he cannot have been unaware of the fact, and we must now ask why.

Two possible lines of interpretation that might be suggested prove, I believe, to be unsatisfactory. First, it is a common modern view that metaphor is an omnipresent feature of all language, and so Aristotle can hardly have failed to use metaphor when explaining metaphor. Although the starting-point of that general view is one with which I have elsewhere expressed a good deal of sympathy, at least insofar as it begins by problematising the literal-metaphorical dichotomy itself,[9] this will hardly do as an explanation of Aristotle's position, for he, after all, was certainly committed to that dichotomy. Besides his account does not just use what we may consider metaphor, but what he certainly has to categorise as 'transfer', *metaphora*, of terms from one domain to another. Given his strictures on such 'transfers', the problem does not go away.

Then it might also be argued that Aristotle had to draw heavily on *metaphora* in this context, since the whole genre of literary criticism and the analysis of language-use itself were, if not invented by Aristotle himself, at least relatively new and under-developed.[10] Yet that is not very satisfactory either. If we consider other fields where Aristotle is certainly an innovator, such as zoology, he is perfectly capable of coining new words and giving them new – strict or proper – definitions, to deal with new subject-matter. Some such coinages are, of course, formed by compounds, where the elements in the compound have regular independent uses. Thus the term Aristotle used for the group of bloodless animals, that correspond roughly to testacea, namely *ostrakoderma*, is compounded of the terms for 'potsherd' (*ostrakon*) and 'skin' (*derma*). Although the two elements are each 'transferred' in one sense (the 'skin' in question is not skin in the strict sense, nor the 'potsherd' a potsherd), the difference is that they are here combined, and the compound, the new coinage, thereby becomes a new lexical entry,

---

[9] Cf. Lloyd 1990, ch. 1.
[10] Claims for his own originality, in the *Rhetoric*, appear in general in I 1 1354a11ff., and in relation to *lexis* in particular at III 1, 1403b15ff. However the study of some aspects of these subjects goes back to the production of manuals, termed *Technai*, by Corax, Teisias and others, beginning in the fifth century, for which Aristotle is one of our principal sources, if generally a highly critical one. We know too that language was an important interest of several of the better known 'sophists', including Protagoras, Hippias, Prodicus and Antiphon, while Plato's own contributions, notably in the *Phaedrus* and *Cratylus*, were in many respects as important as Aristotle's own.

and the proper subject of strict definition. Such a case is analogous to such technical rhetorical terms as *parisōsis* (the balancing of clauses) or *paromoiōsis* (assimilation, of the sounds at the beginning or end of clauses), whether or not these were Aristotle's own invention.

So our problem remains, of saying why Aristotle should have proceeded in the way he did, drawing very heavily on *metaphorai* in his account of *metaphora*. To tackle that, we have to pursue the question we posed earlier, of the status of Aristotle's whole analysis of style, of literary criticism, including language use, in the *Rhetoric* and *Poetics* themselves. The analysis of poetry is not, to be sure, itself poetic. No more does either treatise conform to the model of strict demonstration familiar to us from the *Posterior Analytics*.

But that still leaves open a wide range of options. Rhetoric as *practised*, we are told, has three main branches, deliberative, forensic and epideictic (*Rhet.* 1358b6ff.). Rhetoric as the *theory* is a *technē*, an art, as the titles of works devoted to the subject long before Aristotle had often proclaimed.[11] Aristotle's own *Rhetoric*, his *treatise*, brings that art to a culmination, broadening the scope of the subject-matter, and treating, as he says, of the capacity of discerning the possible means of persuasion on any subject whatsoever (*Rhet.* 1355b26ff.).

Now if Aristotle's own treatment *exemplifies* that, being itself rhetorical in the sense that it *applies* what it teaches about the means of persuasion on any subject-matter (in this case the topic of rhetoric itself), then there would be nothing at all surprising about the *use* of *metaphora* in the account of it. For *metaphora*, as the text teaches us, is an important resource of style, and on this view that would apply to the text of the *Rhetoric* itself, on *metaphora* and throughout. So Aristotle would merely be practising what he preaches, when he advocates *metaphora* as contributing to 'perspicuity, pleasure and a foreign air'.

That offers a clear account of why Aristotle uses *metaphora* so heavily: indeed on this story he might even be said to be obliged to do so, if he is to make the most of the resources of style that metaphor provides. Nevertheless while this account has that very

---

[11] The fragmentary remains of such works are collected in Radermacher 1951.

considerable attraction, it runs into certain difficulties. First it suggests a broader picture of 'rhetoric' than is usually contemplated: second it appears to conflict with some parts of Aristotle's characterisation of it.

Now in *favour* of a broader view of what 'rhetoric' might encompass can be cited certain general remarks that can be read as an indication that, potentially, it has a very wide scope. We are told, for instance, at *Rhetoric* 1354a4ff., that everyone shares, in a manner, in both rhetoric and dialectic: for all try to inquire, to maintain an argument, to defend themselves and to accuse.

Moreover at *Rhetoric* 1404a8ff. he remarks, with regard to style, that it 'necessarily has a small part to play in every mode of instruction (*didaskalia*) for to speak in this or that manner makes a difference to making something clear'. Yet that remark is then immediately qualified: 'but the part is not very great, for all such matters are just for appearance and directed at the listener, and so no one teaches geometry in such a way'. Evidently while the suggestion that style has a part to play in every kind of instruction opens the door wide to the possibility that Aristotle would include all his own treatises as needing thus to pay attention to style, including therefore the proper deployment of *metaphora*, the door gets closed at least insofar as Aristotle himself attempts demonstration in the strict geometrical manner. Yet given that *that* is not in question in *this* treatise, the remark may still be given some self-referential import in this context.

Again while *metaphora*, in particular, is praised for its ability to teach, *mathēsis* (*Rhet.* 1410b10ff.: also *gnōsis*), and again this seems rich in the possibility of application to the Aristotelian Corpus itself, it is not as if the actual examples that he goes on to give, to illustrate the point, are very suggestive of the kind of teaching Aristotle himself wishes to impart. At *Rhetoric* 1410b10ff. the first quotation is from Homer, the verse that calls old age 'stubble', and he continues with a general reference to the 'similes of the poets'. Again at *Rhetoric* 1412a32ff., when a similar point is made, what is learnt is the point of a pun. Even when, most promisingly it might seem, at 1412a9ff., the best metaphors, from 'things that are appropriate but not evident', are compared with the ability, in *philosophy*, to discern similarity even in things that are far apart – there too said to be a sign of an ability to hit the mark (*eustochos*) – the example given is not impressive. True it is ascribed to the philosopher Archytas,[12] but the idea is that there is no difference between an arbitrator and an altar, in that the person wronged takes refuge with both.

So while several texts have the glimmerings of a suggestion that Aristotle appreciated that his own writings, – including the treatises we have,[13] – exemplify the stylistic virtues he identifies in the course of his discussion in the *Rhetoric*, they are no more than glimmerings. The encouragement to see the work as self-referential is minimal, even though he does end with a peroration in his discussion of peroration ('I have spoken, you have heard, you have the case, decide', 1420b3f.). Moreover if we stick to the classification in *Rhetoric* 1358b6ff. (above, p. 215), that specifies the three main species of rhetoric as deliberative, forensic and epideictic, that classification leaves no scope for including Aristotle's own treatise as itself an exemplification of the practice.[14] No more does Aristotle's analysis of the three types of hearer, the member of the assembly, the dicast, the spectator, at whom the techniques of persuasion should be aimed, and whose particular character-traits (including their foibles, not to say their corruption, *mochthēria*) the rhetorician should take into account. Again, among the other points in Aristotle's characterisation of the art that do not sit well with the claim that his own treatise is an example, are that the rhetorician should be able to argue both sides of a case (*Rhet.* 1355a35ff., cf. 1402a26ff.) and that he should be careful to build up a picture of his own character, to portray himself, that is, as a trustworthy person (*Rhet.* II 1, 1377b24ff.).

But if we take the alternative view, and treat Aristotle's own discussion of rhetoric, not as itself rhetoric, but as belonging to philosophy or to dialectic, where does that leave our original question, concerning the heavy use of *metaphora* in Aristotle's own account of that in the *Rhetoric* and *Poetics*?

There would seem to be two main possibilities. The first is that the *Rhetoric* is, like the ethical and political treatises, a contribution to practical philosophy: the second that it should be seen, rather,

---

[12] Otherwise, however, the sources on which Aristotle draws, to illustrate points in his analysis of *metaphora*, are very largely the poets and the orators.

[13] No doubt his lost literary works, modelled in some cases, it seems, on the dialogues of Plato, paid more attention to questions of style.

[14] When written compositions are referred to, for example at *Rhetoric* 1413b4, 8ff., 20ff., he does not specify their subject-matter, but appears to have in mind works that are directly comparable to the orators' public speeches, 1413b14ff.

as an exercise in dialectic, when that is defined, quite generally, as the method by which we shall be able to reason from generally accepted opinions about any problem (*Top.* 100a18ff.).

Now the latter view meets one particular difficulty but itself encounters others. Certainly it is preferable to see the *Rhetoric* as engaged in a study that takes the reputable opinions, the *endoxa*, as starting-point, rather than one that has any pretensions to proceed from the self-evident, indemonstrable premises of the strict demonstrations of the *Posterior Analytics*. Yet if the *Rhetoric* itself aims not at demonstrative certainty, but rather at probability, it still does not exactly fit the notion that dialectic – like rhetoric indeed – argues both sides of a case (*Rhet.* 1355a35ff.).

Moreover if we take the *Rhetoric* as dialectical, it is still doubtful just how far that would help to account for Aristotle's own use of *metaphora* in that treatise. If we review what is said on that subject in the *Topics*, it is true that the examination of likenesses, at least, is there explicitly recognised to be important for securing the genus of a definition and for inductive reasoning.[15] However, references to *metaphora* as such in dialectic are almost uniformly negative. We have commented on the objection at *Topics* 139b34f. (p. 209) that every metaphorical expression is unclear (cf. also 123a33ff.). The remarks at *Topics* 156b10ff. and elsewhere (above p. 208) on using likenesses or *metaphorai* to deceive opponents hardly serve as any kind of justification for Aristotle's own heavy use of *metaphorai* in the *Rhetoric* discussion of *metaphora*.

A more promising hypothesis would be that the *Rhetoric* is not dialectic, but a contribution not, of course, to theoretical, but rather to practical philosophy. True, it would not have the comprehensive goal of the *Nicomachean Ethics*, where that is stated to be to help us to become good (*EN* 1103b26ff.). But it would aim at practical benefits nevertheless, namely to make us better rhetoricians, thus better able to perform our roles as citizens in the matter of public speaking, for instance, as well as in the matter of assessing others' performance in that area.

Thus far the notion of practical philosophy provides a possible framework for understanding this treatise, but it is not as if all our

---

[15] See, for example, *Topics* 105a21ff., 108a7ff., b7ff., and more generally at 114b25ff., 124a15ff., 136b33ff., 138a30ff. A grudging recognition of the value of *metaphora* occurs at *Topics* 140a8ff., where it is compared with what is 'even worse than *metaphora*'; namely expressions that are quite unclear because not based on any similarity at all.

problems disappear on this hypothesis either. The *Nicomachean Ethics* certainly points out expressly that we should not expect exactness, *to akribes*, in the kind of study it undertakes, 1094b11ff., 1104a1ff.: rather what is said is said *tupōi*, in general terms. That clearly allows treatises of practical philosophy to depart from the standards of rigour expected, say, in mathematics. But explicitly to allow inexactness is still not explicitly to allow metaphor. While the ethical treatises frequently refer to expressions used 'metaphorically' or 'according to a likeness' (*kath' homoiotēta*), it is not as if they yield any text that explicitly *legitimates* their use, let alone praises metaphor as having a positive contribution to make to discussion about moral issues.[16] On the contrary, when remarking on a usage that depends on likeness or *metaphora*, Aristotle still generally takes the *strict* sense as the ideal.[17] Even though, noticed or unnoticed, metaphors appear with great frequency, in practice, in his ethics and politics, just as they do in every area of his writing, including his metaphysics and his natural science, his official view, in those treatises of practical philosophy, still appears to be that they are deviant. Just as we found that the analogical use of terms seemed to have, in the final analysis, insufficient theoretical basis in Aristotle's explicit discussions of that question, so too the same may be said with regard to his theory and practice of *metaphora*.

On the self-referential reading of the *Rhetoric*, 'rhetoric', construed broadly, legitimates an extensive deployment of *metaphora*, at least as an ornament of style, and for the purposes of persuasion. On the view that the *Rhetoric* presents a contribution to practical philosophy, it is rather Aristotle's practice that allows for a positive assessment of this mode of language use. But on *either* line of interpretation, Aristotle turns out to be a more positive practitioner of *metaphora* than his critical pronouncements on it lead us to expect.

But of course that still leaves apparent conflicts between, on the one hand, the stipulations of his official theory – not just for the purposes of demonstration, but also for dialectic and even in ethics – and on the other, the actual practice. The official theory stipulates that, for the sake of clarity and the avoidance

---

[16] He does, on occasion, for example at *EN* 1119b3, allow that a *metaphora* may have a point – the transference is 'not badly' done – but, by implication, that is not the norm.

[17] See for example *EN* 1115a14ff., 1138b5ff., 1149a21ff., b31ff., 1167a10f.

of ambiguity, *metaphora* should be avoided. The actual practice exhibits an extensive, even at points an extravagant, use of that very form.

Once again, as in several of our other studies, we have a tension, between an ideal, set out for the sake of the model of philosophical reasoning, and the clear boundaries between different types of inquiry, that it provides, and Aristotle's actual performance, when engaged in a particular study to which, in principle, the ideal should be applicable.

Much is at stake in how we respond to such a tension. How, in general, should we deal with such apparent mismatches between theory and practice?

Some will be inclined not to look beyond the signs that suggest that a verdict of guilty of inconsistency has to be entered. But that provides no more than a judgement, not an explanation.

Are we dealing with development, or at least with changes, in Aristotle's views? On other occasions I have argued that that is possible. But while it may be thought improbable that his views on fundamental topics remained totally unchanged throughout his life, we have, if we wish to argue such a case in a particular instance, always to show grounds both for the earlier, and the later, positions – and for the change between them. Developmental hypotheses, on the question of *metaphora*, would seem more problematic (even) than on other subjects, partly because his actual use of *metaphora* is so widespread and pervasive in just about every field of inquiry he engaged in.

The *metaphora* case bears certain similarities to, but also differences from, some of the others we have investigated in these studies. On the question of demonstration, for instance, Aristotle's own texts show unequivocally, in my view, that he had several theories, several practices, of demonstration: that is, that he thought it could and should take different forms in different inquiries, depending on the nature of the subject-matter, the difficulties of the investigation, the prospects of success.

Again, when dealing with the question of the boundary between plants and animals, he seems prepared to modify his usual, apparently totally confident, one-criterion, stance, in the face of some particularly recalcitrant empirical data.

Again, in the case of concoction, it is the complex and obscure nature of the phenomena he is investigating that seems the main stimulus to his deployment of the concept in ways that outrun (I suggested) the actual theoretical explanations he can give.

Yet, when studying the heavens, it is the reverse: there it is the *lack* of first-hand engagement in the main problems of contemporary astronomical theory that leads to what I diagnosed as a mismatch between his methodological statements and his own contributions to the inquiry.

We have repeatedly found him ready to apply his general ideal differently in different domains or contexts, as in his accounts of the modalities of perception, and in his use of the concept of nature – and the same applies also to such second-order notions as analogy, though that is another case where I suggested that at points his use outstrips the theoretical basis for it he provides. Again, in such an instance as his discussion of spontaneous generation, he is prepared to countenance exceptions that pose serious problems for aspects of his general theories themselves.

No one can doubt the robustness of his theories, not least those that set out methodological prescriptions. But this is often combined with very considerable openmindedness, in his recognition of complexities, variability, and difficulties, whether stemming from the sheer indeterminacy of what is there or from his own uncertainty about it. Such a combination of characteristics is, to be sure, sometimes an uncomfortable one. At the same time it would be absurd to have expected Aristotle to have resolved *all* the difficulties he tackled, ranging, as they did, from such questions as the difference between animals and plants, and why the heavens appear to revolve in one direction rather than the other, all the way to the conditions that satisfactory theories in different domains must meet.

The case we have considered in this chapter resembles that of demonstration most closely, at least in this respect, that Aristotle's view-point seems to vary – and knowingly so, I would again argue – quite radically according to the subject-matter he is dealing with. Engaged in the explicit elaboration of models of reasoning for philosophy, he adopts an attitude that one can only describe as purist. *Metaphora* is not just anathema to syllogistic: it is to be avoided in dialectic as well, except insofar as the dialectician may allow himself tactical gambits to hoodwink partners in debate. Similarly, when in his natural science or his metaphysics, Aristotle comes across rivals or opponents using what he can label *meta-*

phora, he is generally merciless in his condemnation. In both types of case, the positive description of dialectic and the criticisms of opponents, the principal factor that dictates the severity of the tone adopted against *metaphora* is, no doubt, the concern for the self-definition of the philosopher. What we see there is just one reflection of the image Aristotle wants to project of the highest style of philosophising – a topic on which he was locked in battle with friend and foe alike.

Yet when not criticising opponents, and not projecting that image, he makes considerable use of the persuasive resources that *metaphora* can provide. When applying his own skills to the analysis of the arts of rhetoric and poetry, he not only sees that their practitioners draw heavily on *metaphora* – how could he fail to do so? – but he does so himself, making effective use of metaphor to bring out the effectiveness of metaphor itself.

Those who applaud the rigour and purity of the ideal will do nothing but lament this falling off. But those who see the ideal as *just* that, and who take the boundaries between modes of reasoning that it sets up to be themselves idealised abstractions, will be happier to follow the lead that Aristotle himself seems to give: when it comes to a conflict between theory and practice, it is the practice that speaks the louder of the two. That, indeed, has been the guiding thought not just of this, but of our other explorations.

Bibliography

Ackrill, J. L. (1972–3/1979) 'Aristotle's definitions of *psuchē*' (*Proceedings of the Aristotelian Society* 73 (1972–3), 119–33) in Barnes, Schofield, Sorabji 1979: 65–75.
Altman, P. L. and Dittmer, D. S. (1972–4) *Biology Data Book*, 3 vols., 2nd edn (Bethesda, Maryland).
Aubenque, P. ed. (1993) *Aristote politique* (Paris).
Balme, D. M. (1961/1975) 'Aristotle's use of differentiae in zoology' (*Aristote et les problèmes de méthode*, ed. S. Mansion (Louvain, 1961), 195–212) in Barnes, Schofield, Sorabji 1975: 183–93.
  (1962a) '*Genos* and *eidos* in Aristotle's biology', *Classical Quarterly* NS 12: 81–98.
  (1962b) 'Development of biology in Aristotle and Theophrastus: theory of spontaneous generation', *Phronesis* 7: 91–104.
  (1972) *Aristotle's De Partibus Animalium I and De Generatione Animalium I*, Clarendon Aristotle Series (Oxford).
  (1985) 'Aristotle, *Historia Animalium* Book Ten', in *Aristoteles Wirk und Wirkung*, ed. J. Wiesner, vol. 1 (Berlin), 191–206.
  (1987a) 'Aristotle's use of division and differentiae' (revised and expanded version of Balme 1961/1975) in Gotthelf and Lennox 1987: 69–89.
  (1987b) 'Aristotle's biology was not essentialist' (revised and expanded version of article originally published in *Archiv für Geschichte der Philosophie* 62 (1980), 1–12) in Gotthelf and Lennox 1987: 291–312.
  (1991) *Aristotle, History of Animals, Books VII–X* (Loeb Classical Library) (Cambridge, Mass.).
Barker, A. (1981) 'Aristotle on perception and ratios', *Phronesis* 26: 248–66.
Barnes, J. (1969/1975) 'Aristotle's theory of demonstration' (*Phronesis* 14 (1969), 123–52) in Barnes, Schofield, Sorabji 1975: 65–87.
  (1971) 'Homonymy in Aristotle and Speusippus', *Classical Quarterly* NS 21: 65–80.
  (1971–2/1979) 'Aristotle's concept of mind' (*Proceedings of the Aristotelian Society* 72 (1971–2), 101–14), in Barnes, Schofield, Sorabji 1979: 32–41.

(1975) *Aristotle's Posterior Analytics*, Clarendon Aristotle Series (Oxford).
(1981) 'Proof and the syllogism', in Berti 1981: 1–59.
Barnes, J., Schofield, M., Sorabji, R., edd. (1975) *Articles on Aristotle, I Science* (London).
(1979) *Articles on Aristotle, IV Psychology and Aesthetics* (London).
Berti, E., ed. (1981) *Aristotle on Science: the Posterior Analytics* (Padua).
Bolton, R. (1978) 'Aristotle's definitions of the soul: *De Anima* II 1–3', *Phronesis* 23: 258–78.
(1987) 'Definition and scientific method in Aristotle's *Posterior Analytics* and *Generation of Animals*', in Gotthelf and Lennox 1987: 120–66.
(1990) 'The epistemological basis of Aristotelian dialectic', in Devereux and Pellegrin 1990: 185–236.
Broadie, S. (1993) 'Aristotle's perceptual realism', *Southern Journal of Philosophy* 31, Supplement: 137–59. *See also* Waterlow, S.
Burnyeat, M. F. (1992) 'Is an Aristotelian philosophy of mind still credible? A draft', in Nussbaum and Rorty 1992: 15–26.
(1993) 'Aristote voit du rouge et entend un "do"; combien se passe-t-il de choses? Remarques sur *De Anima* II 7–8', *Revue Philosophique de la France et de l'étranger* 183: 263–80.
(1994) 'Enthymeme: Aristotle on the logic of persuasion', in Furley and Nehamas 1994: 3–55.
(forthcoming) 'Signposts in *Metaphysics* Z'
Byl, S. (1968) 'Note sur la place du coeur et la valorisation de la MESOTES dans la biologie d'Aristote', *L'Antiquité classique* 37: 467–76.
(1980) *Recherches sur les grands traités biologiques d'Aristote: sources écrites et préjugés* (Académie Royale de Belgique, Mémoires de la Classe des Lettres, 2nd ser. 64, 3, Brussels).
Charles, D. (1988) 'Aristotle on hypothetical necessity and irreducibility', *Pacific Philosophical Quarterly* 69: 1–53.
(1990) 'Aristotle on meaning, natural kinds and natural history', in Devereux and Pellegrin 1990: 145–67.
(1994) 'Matter and form: unity, persistence and identity', in *Unity, Identity and Explanation in Aristotle's Metaphysics*, edd. T. Scaltsas, D. Charles, M. L. Gill (Oxford), 75–105.
Code, A. (1987) 'Soul as efficient cause in Aristotle's embryology', *Philosophical Topics* 15, 2: 51–9.
(1991) 'Aristotle, Searle, and the mind-body problem', in *John Searle and his Critics*, edd. E. Lepore and R. van Gulick (Oxford), 105–13.
Code, A. and Moravcsik, J. (1992) 'Explaining various forms of living', in Nussbaum and Rorty, 1992: 129–45.
Cohen, S. Marc (1987) 'The credibility of Aristotle's philosophy of mind', in *Aristotle Today: Essays on Aristotle's Ideal of Science*, ed. M. Matthen (Edmonton), 103–25.
Cooper, J. (1987) 'Hypothetical necessity and natural teleology', in Gotthelf and Lennox 1987: 243–74.

(1988) 'Metaphysics in Aristotle's embryology', *Proceedings of the Cambridge Philological Society* NS 34: 14–41.
Derrida, J. (1982) *Margins of Philosophy* (trans. A. Bass of *Marges de la Philosophie*, Paris, 1972) (Brighton).
Devereux, D. T. and Pellegrin, P., edd. (1990) *Biologie, logique et métaphysique chez Aristote* (Paris).
Dicks, D. R. (1970) *Early Greek Astronomy to Aristotle* (London).
Düring, I. (1961) 'Aristotle's method in biology', in *Aristote et les problèmes de méthode*, ed. S. Mansion (Louvain), 213–21.
  (1966) *Aristoteles: Darstellung und Interpretation seines Denkens* (Heidelberg).
Ebert, T. (1983) 'Aristotle on what is done in perceiving', *Zeitschrift für philosophische Forschung* 37: 181–98.
Feyerabend, P. K. (1961) *Knowledge without Foundations* (Oberlin).
  (1975) *Against Method* (London).
Fine, G. (1982) 'Aristotle and the more accurate arguments', in *Language and Logos*, edd. M. Schofield & M. C. Nussbaum (Cambridge), 155–77.
Fortenbaugh, W. W. (1971) 'Aristotle: animals, emotion and moral virtue', *Arethusa* 4: 137–65.
Frede, M. (1985) 'Substance in Aristotle's *Metaphysics*', in Gotthelf, ed. 1985: 17–26.
  (1990) 'The definition of sensible substances in *Metaphysics* Z', in Devereux and Pellegrin 1990: 113–29.
  (1992) 'On Aristotle's conception of the soul', in Nussbaum and Rorty 1992: 93–107.
Frede, M. and Patzig, G. (1988) *Aristoteles Metaphysik Z*, 2 vols. (Munich).
Freeland, C. (1987) 'Aristotle on bodies, matter and potentiality', in Gotthelf and Lennox, 1987: 392–406.
  (1990) 'Scientific explanation and empirical data in Aristotle's *Meteorology*', in Devereux and Pellegrin 1990: 287–320.
  (1992) 'Aristotle on the sense of touch', in Nussbaum & Rorty 1992: 227–48.
Freudenthal, G. (1995) *Aristotle's Theory of Material Substance: heat and pneuma, form and soul* (Oxford).
Furley, D. J. and Nehamas, A., edd. (1994) *Aristotle's Rhetoric: Philosophical Essays* (Princeton).
Furth, M. (1988) *Substance, Form and Psyche* (Cambridge).
Garnsey, P. (forthcoming) *Ancient Theories of Slavery* (Cambridge).
Gotthelf, A. (1985) 'Notes towards a study of substance and essence in Aristotle's *Parts of Animals* II-IV', in Gotthelf ed. 1985: 27–54.
  (1987) 'First principles in Aristotle's *Parts of Animals*', in Gotthelf and Lennox 1987: 167–98.
  (forthcoming) 'First principles in Aristotle's *Parts of Animals*: further reflections' (Seminar to Tokyo Metropolitan University, June 1994).
  ed. (1985) *Aristotle on Nature and Living Things* (Pittsburgh).
Gotthelf, A. and Lennox, J. G., edd. (1987) *Philosophical Issues in Aristotle's Biology* (Cambridge).

Guthrie, W. K. C. (1939) *Aristotle, On the Heavens* (Loeb Classical Library) (Cambridge, Mass.).
Hamlyn, D. W. (1977-8) 'Focal meaning', *Proceedings of the Aristotelian Society* 78: 1-18.
Hanson, N. R. (1973) *Constellations and Conjectures*, ed. W. C. Humphreys (Dordrecht).
Heath, T. E. (1913) *Aristarchus of Samos* (Oxford).
Hesse, M. B. (1980) *Revolutions and Reconstructions in the Philosophy of Science* (Brighton).
Inwood, B. (1979) 'A note on commensurate universals in the *Posterior Analytics*', *Phronesis* 24: 320-9.
Irwin, T. H. (1980-1) 'Homonymy in Aristotle', *Review of Metaphysics* 34: 523-44.
Jaeger, W. (1948) *Aristotle: Fundamentals of the History of his Development*, trans. R. Robinson, 2nd edn (Oxford).
Judson, L. (1991) 'Chance and "Always or for the most part" in Aristotle', in *Aristotle's Physics*, ed. L. Judson (Oxford), 73-99.
Kosman, L. A. (1973) 'Understanding, explanation and insight in the *Posterior Analytics*', in *Exegesis and Argument*, edd. E. N. Lee, A. P. D. Mourelatos, R. N. Rorty (*Phronesis* Suppl. 1) (Assen), 374-92.
(1987) 'Animals and other beings in Aristotle', in Gotthelf and Lennox 1987: 360-91.
(1990) 'Necessity and explanation in Aristotle's *Analytics*', in Devereux and Pellegrin 1990: 349-64.
Kullmann, W. (1974) *Wissenschaft und Methode* (Berlin).
(1980) 'Der Mensch als politisches Lebewesen bei Aristoteles', *Hermes* 108: 419-43.
(forthcoming) 'Requirements for the study of biology according to Aristotle' (Aristotle's Biology Symposium, Bad Homburg 1995).
Labarrière, J. L. (1984) 'Imagination humaine et imagination animale chez Aristote', *Phronesis* 29: 17-49.
(1990) 'De la phronesis animale', in Devereux and Pellegrin 1990: 405-28.
Laks, A. (1994) 'Substitution et connaissance: une interprétation unitaire (ou presque) de la théorie aristotélicienne de la métaphore', in Furley and Nehamas 1994: 283-305.
Lear, J. (1980) *Aristotle and Logical Theory* (Cambridge).
(1988) *Aristotle: The Desire to Understand* (Cambridge).
LeBlanc, C. (1985) *Huai-nan Tzu: Philosophical Synthesis in Early Han Thought* (Hong Kong).
Lennox, J. G. (1980) 'Aristotle on genera, species and "the more and the less"', *Journal of the History of Biology* 13: 321-46.
(1982) 'Teleology, chance and Aristotle's theory of spontaneous generation', *Journal of the History of Philosophy* 20: 219-38.
(1987a) 'Divide and explain: the *Posterior Analytics* in practice', in Gotthelf and Lennox 1987: 90-119.

(1987b) 'Kinds, forms of kinds, and the more and the less in Aristotle's Biology' (revised version of Lennox 1980), in Gotthelf and Lennox 1987: 339-59.
(1991) 'Between data and demonstration: the *Analytics* and the *Historia Animalium*', in *Science and Philosophy in Classical Greece*, ed. A. C. Bowen (New York), 261-94.
(forthcoming) 'Material and formal explanation in Aristotle's *Parts of Animals*' (Oxford Biology Workshop, May 1995).
Leszl, W. (1981) 'Mathematics, axiomatization and the hypotheses', in Berti 1981: 271-328.
Lloyd, G. E. R. (1966) *Polarity and Analogy* (Cambridge).
(1978/1991) 'The empirical basis of the physiology of the *Parva Naturalia* (*Aristotle on Mind and the Senses*, edd. G. E. R. Lloyd and G. E. L. Owen, Cambridge, 1978, 215-39) in Lloyd 1991: ch. 10, 224-47.
(1979) *Magic, Reason and Experience* (Cambridge).
(1983) *Science, Folklore and Ideology* (Cambridge).
(1987) *The Revolutions of Wisdom* (Berkeley).
(1990) *Demystifying Mentalities* (Cambridge).
(1991) *Methods and Problems in Greek Science* (Cambridge).
Major, J. S. (1993) *Heaven and Earth in Early Han Thought* (Albany, N.Y.).
Maula, E., (1974) *Studies in Eudoxus' Homocentric Spheres* (Helsinki).
Mignucci, M. (1975) *L'argomentazione dimostrativa in Aristotele* (Padua).
(1981) 'Ὡς ἐπὶ τὸ πολύ et nécessaire dans la conception aristotélicienne de la science', in Berti 1981, 173-203.
Modrak, D. (1987) *Aristotle: the Power of Perception* (Chicago).
Natali, C. (1990) 'Aristote et la chrématistique', in Patzig 1990, 296-324.
Needham, J. (1956) *Science and Civilization in China, Vol. 2: History of Scientific Thought* (Cambridge).
Neugebauer, O. (1975) *A History of Ancient Mathematical Astronomy*, 3 vols. (Berlin).
Newman, W. L. (1887-1902) *Aristotle, Politics*, 4 vols. (Oxford).
Nussbaum, M. C. (1978) *Aristotle's De Motu Animalium* (Princeton).
Nussbaum, M. C. and Putnam, H. (1992) 'Changing Aristotle's mind', in Nussbaum and Rorty 1992, 27-56.
Nussbaum, M. C. and Rorty, A. O., edd. (1992) *Essays on Aristotle's De Anima* (Oxford).
Owen, G. E. L. (1957/1986) 'A Proof in the *peri ideōn*' (*Journal of Hellenic Studies* 77 (1957), 103-11) in Owen 1986: 165-79.
(1960/1986) 'Logic and metaphysics in some earlier works of Aristotle' (*Aristotle and Plato in the Mid-Fourth Century*, edd. I Düring and G. E. L. Owen (Göteborg, 1960), 163-90), in Owen 1986: 180-99.
(1965/1986) 'Aristotle on the snares of ontology' (*New Essays on Plato and Aristotle*, ed. (J.) R. Bambrough (London, 1965) 69-95) in Owen 1986: 259-78.
(1986) *Logic, Science and Dialectic* (London).
Patzig, G., ed. (1990) *Aristoteles Politik* (Göttingen).

Peck, A. L. (1937) *Aristotle, Parts of Animals* (Loeb Classical Library) (Cambridge, Mass.).
  (1943) *Aristotle, Generation of Animals* (Loeb Classical Library) (Cambridge, Mass.).
  (1965) *Aristotle, Historia Animalium, Books I-III* (Loeb Classical Library) (Cambridge, Mass.).
  (1970) *Aristotle, Historia Animalium, Books IV-VI* (Loeb Classical Library) (Cambridge, Mass.).
Pellegrin, P. (1982) *La Classification des animaux chez Aristote* (Paris).
  (1985) 'Aristotle: a zoology without species', in Gotthelf ed. 1985: 95–115.
  (1986) *Aristotle's Classification of Animals* (revised and trans. A. Preus of Pellegrin 1982) (Berkeley).
  (1987) 'Logical difference and biological difference: the unity of Aristotle's thought', in Gotthelf and Lennox 1987: 313–38.
  (1990) 'Naturalité, excellence, diversité. Politique et biologie chez Aristote', in Patzig ed. 1990: 124–51.
Preus, A. (1975) *Science and Philosophy in Aristotle's Biological Works* (Hildesheim).
Radermacher, L. (1951) *Artium Scriptores* (Österreichische Akademie der Wissenschaften, phil.-hist. Kl., Sitzungsberichte 227,3, Abhandlung, Vienna).
Ricoeur, P. (1977) *The Rule of Metaphor* (trans. R. Czerny with K. McLaughlin and J. Costello of *La Métaphore vive* (Paris, 1975)) (Toronto).
Rorty, R. (1980) *Philosophy and the Mirror of Nature* (Princeton).
Rossitto, C. (1984) 'La dimostrazione dialettica in Aristotele', *La Nottola* 3: 5–40.
Rüsche, F. (1930) *Blut, Leben und Seele* (Paderborn).
Schofield, M. (1990) 'Ideology and philosophy in Aristotle's theory of slavery', in Patzig ed. 1990: 1–27.
Searle, J. R. (1964) 'How to derive "ought" from "is"', *Philosophical Review* 73: 43–58.
Sivin, N. (1996) 'The myth of the naturalists', in *Medicine, Philosophy and Religion in Ancient China: Researches and Reflections* (London) ch. 4.
Smith, R. (1982) 'The syllogism in *Posterior Analytics* I', *Archiv für Geschichte der Philosophie* 64: 113–35.
Solmsen, F. (1929) *Die Entwicklung der aristotelischen Logik und Rhetorik* (Neue Philologische Untersuchungen 4, Berlin).
  (1957) 'The vital heat, the inborn pneuma and the aether', *Journal of Hellenic Studies* 77: 119–23.
  (1960) *Aristotle's System of the Physical World* (Ithaca, N.Y.).
Sorabji, R. (1971/1979) 'Aristotle on demarcating the five senses' (*Philosophical Review* 80 (1971), 55–79) in Barnes, Schofield, Sorabji 1979: 76–92.

(1972) 'Aristotle, mathematics and colour', *Classical Quarterly* NS 22: 293-308.
(1974/1979) 'Body and soul in Aristotle' (*Philosophy* 49 (1974), 63-89) in Barnes, Schofield, Sorabji 1979: 42-64.
(1980) *Necessity, Cause and Blame* (London).
(1991) 'From Aristotle to Brentano: the development of the concept of intentionality', in *Aristotle and the Later Tradition*, edd. H. J. Blumenthal and H. M. Robinson (*Oxford Studies in Ancient Philosophy*, Suppl. vol. 1991), 227-59.
(1992) 'Intentionality and physiological processes; Aristotle's theory of sense-perception', in Nussbaum and Rorty 1992: 195-225.
Tamba-Mecz, I., and Veyne, P. (1979) '*Metaphora* et comparaison selon Aristote', *Revue des Études Grecques* 92: 77-98.
Thompson, D.' A. W. (1910) *Aristotle, Historia Animalium*, in *The Works of Aristotle translated into English*, edd. W. D. Ross et al., vol. 4 (Oxford).
(1947) *A Glossary of Greek Fishes* (London).
Verbeke, G. (1945) *L'Évolution de la doctrine du pneuma* (Paris).
(1978) 'Doctrine du pneuma et entéléchisme chez Aristote', in *Aristotle On Mind and the Senses*, edd. G. E. R. Lloyd and G. E. L. Owen (Cambridge), 215-39.
Wardy, R. B. B. (1990) *The Chain of Change: A Study of Aristotle's Physics VII* (Cambridge).
Waterlow, S. (1982) *Nature, Change and Agency in Aristotle's Physics* (Oxford) See also Broadie, S.
Whiting, J. (1992) 'Living bodies', in Nussbaum and Rorty 1992: 75-91.
Wians, W. (1989) 'Aristotle, demonstration and teaching', *Ancient Philosophy* 9: 245-53.
Wieland, W. (1975) 'Aristotle's physics and the problem of inquiry into principles' (revised version of 'Das Problem der Prinzipienforschung und die aristotelische Physik', *Kant-Studien* 52 (1960-1), 206-19) in Barnes, Schofield, Sorabji 1975: 127-40.
Wildberg, C. (1987) *Philoponus Against Aristotle on the Eternity of the World* (London).
(1988) *John Philoponus' Criticism of Aristotle's Theory of Aether* (Berlin).
Williams, B. (1986) 'Hylomorphism', *Oxford Studies in Ancient Philosophy* (ed. J. Annas) 4: 189-99.
Woods, M. et al. edd. (1984) *Notes on Eta and Theta of Aristotle's Metaphysics recorded by Myles Burnyeat and others* (Oxford).
Wright, L. (1973-4) 'The Astronomy of Eudoxus: Geometry or Physics?', *Studies in History and Philosophy of Science* 4: 165-72.
Xi Zezong (1981) 'Chinese studies in the history of astronomy 1949-1979', *Isis* 72: 456-70.
Xi Zezong and Po Shujen (1966) 'Ancient oriental records of novae and supernovae', *Science* 154: 597-603.

Index of passages referred to

ARISTOTLE
APr. (I 23, 40b23ff.), 13; (41b1ff.), 13; (I 27, 43b33ff.), 144
APo. (I 2), 11–12, 21, 36; (71b16f.), 11; (71b17f.), 10–11; (71b20ff.), 12, 142; (71b22f.), 12, 141; (72b18–20), 11; (I 4), 12, 22; (73b26ff.), 142; (74a10), 13; (75a42ff.), 141; (76a11–15), 143; (I 9, 76a28–30), 15; (76a37ff.), 143; (77a10ff.), 143; (77a26ff.), 143; (I 13), 12; (78a30ff.), 167, 181 n. 21; (78a34f.), 14 n. 10, 21; (78a36), 12; (78b4ff.), 181 n. 21; (78b14), 12; (I 24–26), 13, 22; (86a16ff.), 13, 22; (I 26, 87a1ff.), 19; (87a15f.), 13; (I 27, 87a31ff.), 12, 22; (I 30), 13; (90a7ff.), 181 n. 21; (90a15ff.), 144; (II 6, 92a6ff.), 16; (93a29ff.), 181 n. 21; (93b7ff.), 144; (II 11), 13; (95a14ff.), 181 n. 21; (II 12), 13; (II 13–18), 13–14; (96b36ff.), 141; (97b37ff.), 208; (II 14, 98a1ff.), 141, 145, 148; (98a13ff.), 145–6; (98a20ff.), 145, 149; (II 15), 13; (II 16), 12; (98a37ff.), 181 n. 21; (98a38ff.), 101; (98b36ff.), 101; (II 17, 99a1ff.), 141, 145, 147; (99a15f.), 147–8; (II 17, 99a27ff.), 101; (II 19), 11, 22; (100b11), 11
Top. (I 1), 15–16; (100a18ff.), 218; (100a25ff.), 14; (100a27ff.), 15; (100a29ff.), 15; (105a7ff.), 15; (105a21ff.), 218 n. 15; (106b33ff.), 139 n. 2; (108a7ff.), 218 n. 15; (108b7ff.), 16, 218 n. 15; (108b19), 16; (II 4, 111a14ff.), 16; (111a18f.), 16; (114b25ff.), 218 n. 15; (123a33ff.), 218; (124a15ff.), 218 n. 15; (136b33ff.), 218 n. 15; (138a30ff.), 218 n. 15; (139b34f.), 209, 218; (140a8ff.), 218 n. 15; (141a29ff.), 15; (156b10ff.), 208, 218
SE (167b8ff.), 16; (170a23ff.), 19; (174a37ff.), 208; (176b2off.), 208
Ph. (194a21f.), 201 n. 22; (194a28ff.), 198; (194b16ff.), 52 n. 50; (196b10ff.), 104; (197b18ff.), 104; (198b34ff.), 104; (199a15ff.), 201 n. 22; (199a30ff.), 198; (211a7ff.), 25 n. 13; (233a7), 24; (233b14), 24; (237a35), 24; (238a32), 24; (238b16), 24; (238b26), 24; (240b8), 24; (VII 1–3), 2; (243b7ff.), 97; (259b29ff.), 169; (265b19ff.), 97
Cael. (270b13ff.), 162; (I 4, 270b32ff.), 169; (276b21ff.), 70 n. 2; (277a1ff.), 70 n. 2; (277a8ff.), 70 n. 2; (279b4ff.), 25–6; (279b7ff.), 26; (279b33ff.), 23; (280a5), 23; (II 2, 285b22ff.), 171; (286b1ff.), 169; (II 5, 287b28ff.), 170; (287b34ff.), 182 n. 22; (288a2ff.), 171; (290a26f.), 165; (II 10, 291a29ff.), 170; (II 12, 291b24ff.), 170; (291b35f.), 171–2, 179; (292a1), 172; (292a3ff.), 162; (292a7ff.), 162; (292a14ff.), 170; (293a2ff.), 171
GC (329b26ff.), 96–7; (336a15ff.), 169
Mete. (342b27ff.), 163; (342b29ff.), 165; (343a22ff.), 165; (343a25f.), 165; (343a30ff.), 165; (343b1ff.), 163; (343b4ff.), 163; (343b9ff.), 162; (343b11ff.), 162; (343b18ff.), 163; (343b28ff.), 162; (343b30ff.), 162–3; (I 7, 344a5ff.), 25, 166, 182; (344a35ff.), 166; (344b34ff.), 163; (345a1ff.), 163; (I 8, 345a11ff.), 166; (346b10–15), 166; (357a24ff.), 209; (IV 1, 378b12ff.), 83; (378b20ff.), 96 n. 8; (IV 2, 379b12ff.), 83; (379b18ff.), 84 n. 2; (379b25ff.), 84–5; (379b32), 84; (380a1), 86; (380a6ff.), 84; (380a11f.), 84–5; (380a17f.) 84, 85 n. 3; (380b3ff.), 84; (380b7ff.), 84, 100; (380b13f.), 84; (380b28ff.), 85; (380b30), 85 n. 3; (381a4f.), 85; (381a1of.), 85; (381a23f.), 84; (381b3ff.), 85; (381b6), 201 n. 22; (381b14ff.), 85; (382b13), 70; (383a6f.), 70; (383a13f.), 70; (383b18ff.), 70; (383b2off.), 64 n. 81; (387a32ff.), 152; (389a7ff.), 70; (389a11–19), 70
De An. (402a4ff.), 39; (402b3ff.), 39; (402b25), 24; (403a29ff.), 59 n. 71; (407a25f.), 24; (411a7ff.), 47 n. 25, 121; (411a14ff.), 47 n. 25; (412a2of.), 46; (412a27f.), 40; (412b5f.), 40; (412b6ff.), 40; (413b1f.), 79 n. 12; (413b4f.), 135; (414a2ff.), 135; (414b3), 135; (421a16–20),

231

ARISTOTLE *(cont.)*
135; (421a20f.), 135; (II 11), 130, 132–3; (422b2of.), 133; (422b25ff.), 128; (422b34ff.), 130; (423a13ff.), 44 n. 11; (423b2–6), 131; (423b3), 133; (423b6ff.), 131; (423b15), 131; (423b23–6), 131; (423b26ff.), 44 n. 11, 128; (424a2f.), 129; (424a17ff.), 126; (424a28ff.), 129; (424a32ff.), 133 n. 4; (424b1), 133; (425a9ff.), 200 n. 19; (426b15ff.), 44 n. 11, 131; (429b13f.), 52 n. 49; (432b2of.), 80 n. 13; (435a24ff.), 133; (435b13ff.), 128–9
Sens. (436a6ff.), 40; (436b10ff.), 43 n. 8, 58 n. 69; (438a13ff.), 62 n. 79, 66; (438b3ff.), 62 n. 79; (2, 438b19ff.), 71; (439b25ff.), 71; (440a31ff.), 71; (440b31ff.), 71, 135
Somn. (454a7ff.), 40; (454b24f.), 58 n. 69; (455a27), 135; (456a34ff.), 44 n. 13; (457b1), 129; (458a13ff.), 45 n. 16
Long. (466a18ff.), 45 n. 19; (466b14ff.), 190; (466b21f.), 45 n. 20
Juv. (467b14ff.), 42 n. 5; (467b23ff.), 43 n. 8, 58 n. 69; (467b28ff.), 50 n. 43; (468a13ff.), 50 n. 42, 155; (468b15), 115 n. 14; (469a10ff.), 47 n. 27; (469a12ff.), 47 n. 28, 128; (469a14), 128; (469a18ff.), 58 n. 69; (469a2of.), 128; (469b3ff.), 58 n. 69; (469b8ff.), 46 n. 22; (470a19ff.), 46 n. 22
Resp. (471b19ff.), 115 n. 14; (473a9f.), 46 n. 22; (474b3ff.), 44 n. 13; (477a16ff.), 47 n. 26; (477a18ff.), 45 n. 16; (479a3ff.), 115 n. 14
HA (486b19ff.), 141, 149, 152; (487a11ff.), 49, 186 n. 3; (487b6), 77 n. 10; (487b9), 75 n. 9; (487b10f.), 75; (487b15), 77 n. 10; (487b22f.), 152; (487b25ff.), 80 n. 13; (487b34ff.), 49 n. 41; (488a3ff.), 186; (488a7ff.), 186; (488a9ff.), 186; (488a10ff.), 190; (488b12ff.), 186 n. 3; (488b13ff.), 49 n. 40, 186 n. 5; (488b16f.), 186 n. 5; (488b29ff.), 49; (489a17f.), 135; (489a20ff.), 45 n. 20, 154; (489a21ff.), 155; (491a9f.), 55 n. 58; (491a11ff.), 24; (491a14ff.), 57 n. 64, 72; (494a26ff.), 190 n. 9; (494a33ff.), 190 n. 9; (494b16–18), 135; (497b32ff.), 149; (511b4), 155; (515b27ff.), 155; (516b3ff.), 150; (516b12ff.), 150; (516b31f.), 150; (519a27ff.), 115 n. 14; (521a9f.), 87; (521a17ff.), 87; (521b2f.), 87, 155; (IV 1–7), 50, 154; (523b2ff.), 43 n. 9; (523b5ff.), 43 n. 9; (523b15f.), 132, 153; (524b14ff.), 155; (527a17ff.), 155; (527b28f.), 155; (528a6f.), 135, 154; (528a8f.), 134; (529a11f.), 155; (531a8ff.), 76; (531a11ff.), 76; (531a14f.), 76; (531a17f.), 76; (531a19ff.), 76; (531a27ff.), 76; (531a31ff.), 77; (531b1f.), 77; (531b8ff.), 77; (531b10ff.), 77; (531b30ff.), 156; (532a31ff.), 132, 153; (IV 8, 532b29–535a27), 79 n. 12; (533a2ff.), 81, 200 n. 19; (534b15ff.), 135; (535a4ff.), 135; (535a14ff.), 78; (537b22ff.), 151; (537b24f.), 151; (538a8ff.), 92; (538a18f.), 151; (538a25ff.), 190; (V 1, 539a22ff.), 105; (539b2–7), 105; (539b7ff.), 105; (546b15ff.), 114; (546b26ff.), 115; (V 15, 547b15ff.), 77; (547b18ff.), 117; (548a4ff.), 78; (548a5f.), 78; (548a10f.), 77 n. 10; (548a24ff.), 77; (548b5ff.), 76; (548b10ff.), 75; (548b17f.), 76; (549a7ff.), 76; (V 19), 114; (550b32ff.), 94 n. 6, 116; (551a2ff.), 95 n. 7; (551a13ff.), 80 n. 13, 114, 117; (551a18), 116; (551a24ff.), 116; (551a29ff.), 114; (551b9ff.), 114; (551b16ff.), 117; (552a15), 117; (552a24ff.), 80 n. 13; (552a29), 117; (552a31ff.), 117; (552b4ff.), 117; (552b6ff.), 94 n. 6, 116; (552b10ff.), 94 n. 6, 116; (555b18–556a7), 94; (557b1ff.), 94 n. 6, 117; (557b6), 117 n. 18; (557b8ff.), 117; (557b12ff.), 114; (557b23f.), 116; (557b24f.), 114; (557b25ff.), 86; (VI 3, 561a4ff.), 156; (562b17ff.), 93; (565b22f.), 93; (568a4–8), 134; (VI 15), 114; (569a13ff.), 115; (570a3ff.), 115; (570a15ff.), 115; (573b34ff.), 94 n. 5; (588a17ff.), 186 n. 3; (588a28ff.), 149; (588b4–589a2), 73–9; (588b4ff.), 47 n. 25, 67, 74; (588b5f.), 74; (588b6), 75; (588b7), 74; (588b8), 73; (588b9), 73; (588b10–12), 73–4; (588b12ff.), 74, 78; (588b13), 75; (588b17f.), 73–4, 79; (588b19f.), 76–7; (588b20f.), 75; (588b21), 73; (588b22), 73; (588b24ff.), 73, 79; (588b26f.), 73–4, 79; (588b31), 73; (VIII 2, 589a10–590a18), 73; (590a28ff.), 77; (590b10), 191 n. 10; (594a4f.), 191 n. 10; (596b10f.), 191 n. 10; (606a8), 112; (IX), 49, 186 n. 3; (608a33ff.), 190 n. 8; (X), 49 n. 39
PA (I 1), 28–34, 36; (639b1ff.), 57 n. 61; (I 1, 639b5ff.), 29; (639b11ff.), 29; (639b21ff.), 29–30; (639b30–640a8), 29, 99 n. 9; (640a3f.), 31; (640a6f.), 30; (640a28f.), 105; (641a1ff.), 58 n. 67; (641a17ff.), 39; (641a18ff.), 58; (641a29ff.), 40; (641a34ff.), 39; (642a9ff.), 58 n. 67; (I 2–4), 55, 71; (642b10ff.), 55; (642b15), 57 n. 63; (642b20ff.), 55; (642b31ff.), 56; (643a1ff.), 59; (643a3ff.), 60; (643a7ff.), 55; (643a13ff.), 56; (643a17), 55; (643a24ff.), 40 n. 3; (643a25), 62; (643a27f.), 55; (643b1ff.), 72, 108 n. 7; (643b10ff.), 55; (643b12), 55; (644a10ff.), 55; (I 4, 644a19ff.), 71; (644b7ff.), 57 n. 64, 72; (I 5, 644b22ff.), 161; (644b31ff.), 161; (645a30ff.), 40, 55; (645a35ff.), 55 n. 58, 61; (645b6ff.), 152; (645b14ff.), 58; (645b19),

Index of passages referred to 233

40; (646b10ff.), 58 n. 67; (647a16ff.), 128; (647a19ff.), 44 n. 11, 130; (647a21), 79 n. 12; (647a24ff.), 47 n. 27; (647b4ff.), 87; (647b26f.), 45 n. 20; (647b29ff.), 60 n. 75; (647b31ff.), 45 n. 16; (648a2ff.), 45 n.16; (648a5ff.), 186 n. 4; (648a13ff.), 60 n. 75; (648a19ff.), 152; (648a28ff.), 92; (II 2, 648a36ff.), 96; (648b4ff.), 45 n. 18; (648b12ff.), 129; (648b26ff.), 129; (648b30ff.), 129; (649a10f.), 129; (649b2ff.), 129; (II 3, 649b10ff.), 96; (650a3ff.), 86; (650a10ff.), 86; (650a14), 46 n. 22; (650a20ff.), 86; (650a28ff.), 86; (650a34f.), 86; (650b2ff.), 44; (650b8ff.), 87; (650b12f.), 44; (650b18ff.), 45 n. 16; (651a12ff.), 44, 49 n. 40; (651a17ff.), 87, 155; (651a20ff.), 45 n. 17, 87; (651b20ff.), 45 n. 17; (651b28ff.), 87; (652a6ff.), 44 n. 13, 87; (652a7ff.), 87, 98; (652a24ff.), 47 n. 29, 88; (652b6ff.), 47 n. 29; (652b7ff.), 46 n. 23; (653b19ff.), 43; (653b21f.), 128; (653b22ff.), 57, 58 n. 69; (653b24–30), 130; (653b30ff.), 43; (653b35ff.), 43 n. 10, 150; (653b36ff.), 43 n. 9; (654a5ff.), 133; (654a12ff.), 132; (654a22ff.), 132 n. 3; (654a26ff.), 132, 153; (654a32ff.), 128; (654b27ff.), 43 n. 10; (655a32ff.), 150; (655b20ff.), 58 n. 67; (655b29ff.), 50 n. 42, 155; (655b32ff.), 86; (656a10ff.), 190; (656a15ff.), 47 n. 30; (656a19ff.), 47 n. 29; (656a29ff.), 128; (656a37), 128; (656b16ff.), 128; (656b18f.), 128; (656b35f.), 44 n. 11; (657a22ff.), 81; (659b30ff.), 42 n. 4; (660a11ff.), 135; (660a19ff.), 42 n. 4; (660a22ff.), 42 n. 4; (660a35ff.), 42 n. 4; (661b1ff.), 42 n. 4; (661b13ff.), 42 n. 4; (662b24ff.), 151; (663b35ff.), 145; (665b5f.), 87; (665b14ff.), 87; (665b18ff.), 48 n. 36; (665b22ff.), 48 n. 35; (666a6ff.), 87; (666a34), 58 n. 69; (666b6ff.), 48 n. 36; (667a11ff.), 45 n. 16; (667b21ff.), 157; (668a4ff.), 44 n. 15; (668a5f.), 44 n. 12; (668a11f.), 44 n. 15; (668a13ff.), 44 n. 15; (668a16ff.), 44 n. 15; (668a25ff.), 44; (668b1ff.), 88; (668b8ff.), 87; (668b11ff.), 89; (668b24ff.), 44; (668b33ff.), 47 n. 29;(670a8ff.), 44 n. 14; (670a20f.), 87; (670a22ff.), 87; (670a27), 87; (672a1ff.), 87; (672a20ff.), 87; (672b14ff.), 48 n. 31; (672b31ff.), 48 n. 31; (673b26ff.), 45 n. 17; (674a6ff.), 45 n. 17; (III 14), 89; (674a28–31), 89; (674a31ff.), 145; (674b4ff.), 87; (674b26ff.), 90; (674b34ff.), 90; (675a27ff.), 89; (676b22ff.), 47 n. 30; (677a15ff.), 47; (IV 3, 677b14ff.), 88; (678a6ff.), 44 n. 13; (678a26ff.), 57 n. 61, 154; (678a33ff.), 59; (678a36ff.), 154; (679a25f.), 60 n. 74, 154; (679b10ff.), 155; (679b34), 135, 154; (680a4ff.), 76; (680a5ff.), 154; (680a11f.), 154; (680a20f.), 155; (681a2ff.), 90; (681a9–b12), 67, 73, 76; (681a11f.), 75; (681a12ff.), 47 n. 25, 74; (681a14), 73; (681a15ff.), 73, 75–6; (681a17ff.), 76, 79; (681a19f.), 77; (681a20ff.), 73, 79; (681a24), 80; (681a25ff.), 76; (681a27ff.), 73–6; (681a31ff.), 74, 76; (681a33f.), 79; (681a35ff.), 74–5, 77; (681b2ff.), 77; (681b5ff.), 77, 80–1; (681b7ff.), 73–4; (681b26ff.), 150, 155; (681b33ff.), 50 n. 43, 156; (682a1ff.), 50 n. 44, 156; (682a6ff.), 157; (682b2off.), 157; (682b29ff.), 80 n. 13; (683b5), 80 n. 13; (683b8ff.), 80 n. 13; (683b9ff.), 77; (683b17), 78; (683b18ff.), 48 n. 32, 76; (684a1ff.), 134; (684a33ff.), 152; (686a8ff.), 47 n. 30; (686a27ff.), 58 n. 69; (686b2ff.), 120 n. 23; (686b20ff.), 120 n. 23; (686b31ff.), 48 n. 32; (687a7ff.), 58 n. 67; (689b25ff.), 120 n. 23; (692b8ff.), 58 n. 65; (692b22ff.), 58 n. 64; (693a10ff.), 58 n. 64; (693a26ff.), 200 n. 20; (693b2ff.), 60; (693b6ff.), 59; (693b13), 58, 80 n. 13; (694a1ff.), 58 n. 64; (694b12ff.), 58 n. 64; (694b13ff.), 58 n. 67; (695a8ff.), 120 n. 23; (695b17ff.), 59; (695b20), 59; (697b7ff.), 152; (697b14ff.), 80 n. 13

MA (698a5ff.), 57 n. 61; (703a34ff.), 42 n. 5

IA (705a32ff.), 48 n. 32; (705b6), 48 n. 32; (705b8ff.), 48 n. 34; (705b29ff.), 48 n. 33; (706a16ff.), 190 n. 9; (706a20ff.), 190 n. 9; (706b9ff.), 190 n. 9; (706b12f.), 48 n. 35; (707a24ff.), 157; (707b2ff.), 157; (708b4ff.), 115 n. 14; (712b22ff.), 200 n. 20; (714a21ff.), 58 n. 64; (714b8ff.), 81

GA (715a24f.), 94; (715b16ff.), 58 n. 69, 91; (715b19ff.), 151; (715b21ff.), 85; (715b26ff.), 94; (716a4ff.), 91; (717a21f.); 79; (719a33ff.), 92; (719a35ff.), 91; (725a3ff.), 86, 91; (725b21f.), 89, 91; (726a3ff.), 91; (726a6ff.), 86; (726a26ff.), 91; (726b1ff.), 44 n. 13, 86; (726b3ff.), 155; (726b5f.), 87, 91; (726b9ff.), 45 n. 17; (726b22ff.), 40 n. 3, 42 n.5; (726b30ff.), 92; (727a2ff.), 42 n. 7; (727a3f.), 149, 153; (728a17ff.), 91; (728a20f.), 155; (728a21ff.), 89; (728a26ff.), 42 n. 7, 91, 153; (729a16ff.), 92, 94; (730a15ff.), 92; (730a29ff.), 93; (731a1ff.), 91, 151; (731a24ff.), 58 n. 69; (731b4f.), 58 n. 69; (731b10f.), 151; (732a12f.), 79 n. 12; (732a18f.), 46 n. 22; (732b11ff.), 105; (732b15ff.), 57; (732b31ff.), 47 n. 26; (732b32), 46 n. 22; (733a11f.), 45; (733a33ff.), 47 n. 26; (734b24ff.), 40 n. 3, 42 n. 5; (735a4ff.), 42 n. 6; (735a6ff.), 40 n. 3, 42 n. 5; (735a16ff.), 42 n. 6; (735b19ff.), 64 n. 81; (735b33), 64 n. 81; (736a1), 64 n. 81; (736a9), 64 n. 81; (736a30f.), 58 n. 69; (736a35ff.), 42 n. 6; (736b1ff.), 42 n. 6; (736b5ff.), 42

234　　　　　　　*Index of passages referred to*

ARISTOTLE (*cont.*)
n. 6; (736b8ff.), 42; (736b34ff.), 46 n. 23, 149; (737a3ff.), 95, 119 n. 21; (737a7ff.), 42 n. 6; (737a12ff.), 92; (737a27ff.), 42 n. 7, 91, 153, 199 n. 17; (737b26ff.), 64 n. 81; (738b25ff.), 42; (739a11), 46 n. 22; (739b22f.), 92; (740a21f.), 44 n. 13; (740b3), 44 n. 13; (741a3f.), 151; (741a9ff.), 40 n. 3, 58 n. 69; (741a23ff.), 42 n. 5; (742b17-35), 27; (742b32f.), 27; (742b33ff.), 27; (743a1ff.), 44 n. 14; (743a26-34), 47; (743a35f.), 95, 117; (743b29ff.), 47 n. 29; (744a16ff.), 88, 93; (744b1ff.), 88, 93; (747a28), 26; (747b27ff.), 26; (748a7ff.), 27; (748a12), 27; (750b24ff.), 90, 93; (751a34ff.), 44 n. 12; (752a2f.), 46 n. 22; (752b15ff.), 93; (752b29ff.), 93; (752b32ff.), 93; (753a5ff.), 94; (753a7), 93; (753a21ff.), 94; (753a34ff.), 93; (753b2), 93; (753b9f.), 93; (755a20), 46 n. 22; (756b5ff.), 112; (756b10ff.), 92; (756b27ff.), 92; (757b15ff.), 42 n. 6; (III 9), 114; (758b28ff.), 116; (759a20), 190; (760a12), 149; (III 11), 114, 120–1, 123; (761a13ff.), 123 n. 25; (761a26ff.), 149; (761b11), 64 n. 81; (761b13–22), 123 n. 25; (761b23ff.), 119 n. 21; (761b35), 116 n. 16; (762a2ff.), 116 n. 16; (762a8ff.), 114; (762a9ff.), 118; (762a13ff.), 94, 118; (762a18ff.), 46 n. 23, 64 n. 81, 119 n. 21, 120; (762a20), 46 n. 22; (762a24ff.), 120, 123 n.25; (762a32ff.), 114; (762a35ff.), 117; (762b4ff.), 117; (762b9ff.), 120, 123; (762b12ff.), 95; (762b14ff.), 95, 117, 119 n. 21; (762b26ff.), 115; (763a30ff.), 115 n. 15; (763b1ff.), 115 n. 15; (IV 1–3), 123 n. 26; (765b8ff.), 91–2; (766a35), 46 n. 22; (766b12–16), 91, 153; (IV 2–3), 94; (766b34), 46 n. 22, 94 n. 5; (IV 3), 42 n. 7; (767b15ff.), 94; (768a10ff.), 42 n. 7; (768b25ff.), 94; (774a2f.), 91; (774b31ff.), 115 n. 14; (775a4ff.), 93; (775a14ff.), 91, 199 n. 17; (775a16ff.), 93; (775a20f.), 93; (776a25ff.), 90; (778b32ff.), 58 n. 69; (780b6ff.), 89; (783a19ff.), 90; (783b10ff.), 102 n. 11; (783b30), 46 n. 22; (784a32ff.), 45 n. 18; (784a34ff.), 89; (784a35f.), 46 n. 21; (784b5), 46 n. 21; (784b8ff.), 151; (784b11ff.), 89; (784b19ff.), 151; (784b26), 46 n. 22; (784b32ff.), 89; (786a11), 46 n. 22; (786a15ff.), 90; (786a20f.), 46 n. 21; (788b3ff.), 42 n. 4

Metaph. (982a25ff.), 22; (982b12ff.), 104; (982b18ff.), 112 n. 9; (990a24), 20; (991a20ff.), 209; (997a31), 20; (1000a9ff.), 112; (1000a19f.), 20; (1003a33ff.), 139 n. 2; (1003b12ff.), 142; (1006a5ff.), 20; (1006a10f.), 20; (1006a11f.), 20; (1006a17f.), 20–1; (1006a21), 20; (1006a24), 21; (1011a13), 20; (1012b34ff.), 28; (1013a15), 20; (1015b6ff.), 21; (1016b31ff.), 140; (E 1, 1025b7), 21–2, 36; (1025b10ff.), 21; (1025b13), 21–2; (1025b14ff.), 22; (1025b18ff.), 21; (1025b24ff.), 30; (1025b30ff.), 52 n. 48, 69; (1025b34ff.), 52 n. 49; (1026b27ff.), 22; (1027a8ff.), 22; (1030a35ff.), 139 n. 2; (Z 7, 1032a28ff.), 119 n. 21; (1032a30ff.), 119 n. 21; (1032b4ff.), 119 n. 21; (1032b21ff.), 119 n. 21; (Z 9, 1034a9ff.), 119 n. 21; (1034a26ff.), 119 n. 21; (1034b4ff.), 119, 123; (1034b5), 119 n. 21; (1034b16ff.), 119; (Z 10–11), 53; (1035a18ff.), 62; (1035b16ff.), 53; (1036b22ff.), 53; (1036b28ff.), 53; (1037a18–20), 53 n. 53; (1037a21–b7), 53 n. 53; (1037a21ff.), 54, 68; (1037a26ff.), 59; (1037a27ff.), 54; (Z 12, 1037b10ff.), 65 n. 83; (Z 13, 1038b1ff.), 53 n. 53; (1039a14ff.), 53 n. 53; (1039a20ff.), 53 n. 53; (1039b27–1040a5), 20; (Z 17, 1040b5ff.), 70; (1040b8ff.), 70; (1044a9ff.), 57 n. 64; (1044a29), 62 n. 80; (1045a7ff.), 65 n. 83; (1045a23ff.), 65 n. 83; (Θ 6, 1048a35ff.), 140, 142, 144; (1060b37ff.), 139 n. 2; (1062a3ff.), 21; (1062a5ff.), 21; (1062a9ff.), 21; (1062a30ff.), 21; (1064a21ff.), 52 n. 48; (1070a3iff.), 139; (1070b17ff.), 140; (1071b26ff.), 112; (Λ 8, 1073a14ff.), 182 n. 22; (1073a22), 182 n. 22; (1073b3ff.), 174; (1073b4ff.), 181; (1073b6ff.), 181; (1073b8ff.), 174, 180; (1073b10ff.), 164, 174; (1073b11ff.), 175; (1073b13ff.), 174, 180; (1073b14), 174 n. 13; (1073b16f.), 174; (1073b38ff.), 174, 177; (1074a6ff.), 177; (1074a11f.), 160 n. 1; (1074a12ff.), 176–7, 179, 182; (1074a13f.), 160 n. 2; (1074a14ff.), 174; (1074a16f.), 164, 182; (1074a38ff.), 112 n. 9; (1074b4), 112 n. 9; (1078a9ff.), 22; (1079b24ff.), 209; (1087b21), 20

EN (1094b1ff.), 219; (1094b25ff.), 19; (1096b27ff.), 139; (1098b1ff.), 19; (1098b27ff.), 19; (1103b26ff.), 218; (1104a1ff.), 219; (1115a14f.), 219 n. 17; (1116b24f.), 186 n. 5; (1116b30ff.), 186 n. 5; (1118a23ff.), 186 n. 5; (1119b3), 219 n. 16; (1126b3ff.), 126; (1138b5ff.), 219 n. 17; (1139b18ff.), 18; (1139b26ff.), 18; (1139b31f.), 18; (1140b31–1141a8), 18; (1143a35ff.), 18; (1143b10ff.), 18; (1144b1ff.), 202 n. 25; (1145b2ff.), 18, 25 n. 13; (1145b7), 18; (1149a21ff.), 219 n. 17; (1149b31ff.), 219 n. 17; (1167a10f.), 219 n. 17; (1170a16), 79 n. 12; (1179b20ff.), 197

EE (1214a32ff.), 19; (I 3, 1214b27ff.), 19; (1215a7ff.), 19, 26

Pol. (1252a28ff.), 185, 188–9; (1252b28ff.), 185; (1253a27ff.), 188; (1253a29ff.), 188;

(1253b20ff.), 189; (1254a28ff.), 189; (1254a36ff.), 199; (1254b4ff.), 189; (1254b13ff.), 189; (1255a4ff.), 189; (1255a31ff.), 189; (1255b4ff.), 189; (1255b37ff.), 193; (1256a1ff.), 184, 191; (1256a19ff.), 191; (1256a26f.), 191; (1256a36), 192; (1256a38ff.), 191; (1256b5), 192; (1256b7ff.), 192; (1256b10ff.), 192; (1256b23–6), 192; (1256b26ff.), 192, 194; (1256b27ff.), 193; (1256b30f.), 194; (1256b40ff.), 194; (1257a3ff.), 192 n. 11, 193; (1257a29), 193; (1257a30), 193; (1257a33f.), 195; (1257b19ff.), 193; (1257b30ff.), 193; (1258a5f.), 193; (1258a14ff.), 193; (1258a18), 194; (1258a27ff.), 194; (1258a38ff.), 194; (1258b1), 194; (1258b27ff.), 193 n. 13, 194; (1265a10ff.), 187; (1284b15ff.), 196; (1287b39ff.), 196; (1290b21ff.), 195; (1290b25ff.), 57 n. 62; (1309b23ff.), 196; (1320b33ff.), 196; (1326a7), 188; (1326a35ff.), 188; (1326b5ff.), 187; (1326b7ff.), 187; (1328b24ff.), 195 n. 15; (1332a39ff.), 197; (1332b35ff.), 189; (1333a35), 192; (1333b38), 192–3; (1334a2), 193

Rh. (1354a4ff.), 16, 216; (1354a11ff.), 214 n. 10; (1355a4ff.), 17; (1355a35ff.), 217; (1355b26ff.), 215; (1358b6ff.), 215, 217; (1368a32f.), 17; (II 1, 1377b24ff.), 217; (II 1, 1378a6ff.), 17; (1394b8ff.), 17; (1402a26ff.), 217; (II 25, 1403a10ff.), 17; (III 1, 1403b11ff.), 17; (III 1, 1403b15ff.), 211 n. 7, 214 n. 10; (1404a8ff.), 216; (1404b4ff.), 213; (1404b33ff.), 210; (1404b36), 212; (1405a4ff.), 210; (1405a8ff.), 212; (1405a9f.), 211; (1405b11ff.), 207; (1406a14ff.), 212; (1406a32), 212 n. 8; (1406b4ff.), 212; (1406b10f.), 212 n. 8; (1406b20), 207; (1406b21ff.), 207; (1406b24ff.), 210; (1407a13ff.), 138, 207 n. 2; (1410b10ff.), 216; (1410b17f.), 207; (1410b31ff.), 213; (1410b33ff.), 213; (1410b36ff.), 138; (1411a26ff.), 213; (1411a35ff.), 213; (1411b4ff.), 213; (1411b24ff.), 213; (1411b26), 213; (1411b28), 213; (1411b29), 213; (1411b30), 213; (1411b33), 213; (1412a2), 213; (1412a3), 213; (1412a4), 213; (1412a8), 213; (1412a9ff.), 212–13, 216; (1412a32ff.), 216; (1412b32ff.), 207; (1412b34ff.), 138; (1413a13f.), 207; (1413b4), 217 n. 14; (1413b8ff.), 217 n. 14; (1413b14ff.), 217 n. 14; (1413b20ff.), 217 n. 14; (1414a30), 24; (1417b21), 17; (1417b32ff.), 18; (1418a26), 24; (1420b3f.), 217

Po. (1447b17ff.), 210; (1450b13ff.), 211 n. 7; (1457b6ff.), 206, 211–12; (22, 1458a18ff.), 213; (1458a22f.), 212; (1459a6ff.), 211

PSEUDO-ARISTOTLE
De Mundo (392a26), 167 n.9

Ars Eudoxi (5.10), 167 n. 9; (22.1–23.14), 177 n. 16

CLEOMEDES
De Motu Circulari (190.17ff.), 168 n. 11

PLATO
Lg. (822a), 169, 175; (889a), 201 n. 21; (890d), 201 n. 21; (892b), 201 n. 21; (896e ff.), 201 n. 21
Prt. (320c ff.), 186 n. 2
R. (530b), 166; (616e f.), 166
Ti. (36d5), 172; (39d), 175; (40d2ff.), 169; (71a ff.), 47 n. 30; (77b), 79 n. 12

PLINY
HN (II 95), 164

PTOLEMY
Syntaxis (VII 2), 164

SIMPLICIUS
In Cael. (88.31ff.), 167 n. 7; (493.11ff.), 177 n. 17; (495.26ff.), 172 n. 12; (503.10ff.), 160 n. 2; (504.17ff.), 168 n. 10; (504.25–505.19), 168 n. 11

General index

acquisition, modes of, 191–4
actuality/potentiality, 41, 65, 96, 118–19, 139–40, 142–4, 146, 158
air, 45, 64, 71
aither, 166, 181
amber, 70
ambiguity, 73, 139, 189
analogy, 37, 42–3, 46 n. 23, 50, 60, 71, 87, 91–2, 116, 120–1, 126, 129, 131–3, 135, ch. 7, 189, 191, 195–6, 200–1, 209, 219, 221
anthropocentricity, 204
anthropomorphism, 112 n. 9
aquatic, 73
Archytas, 217
aristocracy, 196
arithmetic, 143, 181
art, artefacts, 29–32, 61–3, 65, 85, 149, 188, 201, 215
ash, 87
astronomy, 13, 101, ch. 8, 221
atomists, 63–5
audience, 18, 217
axiomatics, 8, 12, 20, 36, 100, 102, 143

Babylonia, 162
balance, 47, 85, 89, 94, *see also* proportion
Balme, D. M., 56
bat, 152
baths, 84
beak, 58 n. 64, 141
bear, 190
bees, 114, 149, 186, 190
bile, 47
bipedality, 57, 60
birds, 41, 55, 58 n. 64, 59–60, 71, 80 n. 13, 89, 93, 105, 109, 121, 141–2, 145, 191, 200
bladder, 42, 154
blind, 81
blood, 43–5, 49, 51, 60, 62, 86–9, 91, 97–100, 141, 149–50, 152–5
blooded/bloodless animals, 43, 50, 59–61, 76, 80, 86, 127, 132, 135, 137, 150, 154–6, 214
blood-vessels, 43–5, 86, 89, 128
body, 39–40, 42, 44, 51, 56, 59, 128, 132, 189, 198
boiling, 83–5, 101
boils, 84, 96
bones, 43–4, 62–3, 70, 128, 132–4, 137, 141, 145, 148–50, 152–3
botany, 13, 145, *see also* plants
brain, 47, 50, 88–9, 93, 97, 128
brigandage, 192
brightness (of heavenly bodies), 162, 164, 167–8, 180
Buddhism, 106, 109
burning, 87, 129
Burnyeat, Myles, 17, 63–4, 118 n. 20, 126–7
bussos, 77–8
butterfly, 80 n. 13, 114, 116–17

calamaries, 132
Callippus, 160, 168, 172–3, 176–80
camel, 89, 145
caprifig, 85
carnivores, 191
cartilage, 43, 150
catarrhs, 84
caterpillar, 114, 116–17
causation, 11–12, 21, 23, 27, 35, 51, 116, 122, 139–40, 143–4, 146–7, 182
cephalopods, 43, 50, 132, 150, 154–6
chance, 104
character, 18, 44–5, 49, 60, 108, 149, 154, 186, 217
China, 106–11, 113, 124–5, 164 n. 4, 165
choice, moral, 186
chrysalis, 114, 116
circular motion, 169, 181
citizens, 187
classification, of animals, 56, 72, 81, 107, 195
of political constitutions, 57 n. 62, 195, 198

236

General index

claw, 141, 149, 152
clay, 84
coagulation, 13, 101–2
cocoon, 114
coinage, 193, 195
coinage of terms, 150, 214
colour, 71, 89, 126, 140, 151, 164, 166–7, 181
comets, 25, 162–6, 181
commensurability, 19
commensurate universal, 12, 142
comparisons, 200, 206–7, 209
composite wholes, 52–4, 59–63, 69
concentric spheres, 167–9, 172, 178–80
concoction, 42 n. 7, 47, 70, ch. 4, 117–18, 153, 155, 220
concords, 71
conjunctions, of heavenly bodies, 162–3, 165
contact, 131, 133
continuity, in nature, 74
contradiction, law of non-, 20, 72–3, 143
control centre, 50, 128, 155–6
convention, custom, 110, 188, 198, 200–1, *see also* nomos
cooking, 86, 102
copulation, 105, 114–15, 151
correlations, 51, 57, 108
cosmology, 23, 25, 112, 169, 175, 180, 204
courage, 44, 186, 190
crabs, 48, 134, 155
crayfish, 48, 155
creative potential of nature, 120–1
crustacea, 43, 132–5, 154, 191
Ctesias, 112
curdling, 92
cuttlefish, 132, 145

death, 45, 64, 126, 128
debate, 208
deciduousness, 13, 101–2
deduction, 9, 12–14, 23, 35–7, 141–2, 208
  of antecedents, 32–4, 36–7
defence, 42, *see also* self-preservation
definition, 8–9, 12, 24, 37–9, 41, 52–9, 61, 64, 68–9, 83, 96–7, 100–2, 116, 138, 140–4, 147–8, 153–4, 208–9, 212, 214–15, 218
deformity, 68, 81, 91, 93, 120, 152, 199–200, 203
deliberative oratory, 18, 215, 217
democracy, 196
Democritus, 26–8, 163
demonstration, ch. 1, 51, 83, 99 n. 9, 100, 102, 141, 143–5, 147–8, 166, 181–2, 208–9, 215–16, 218, 220–1

Descartes, 63–4
desire, 40
detachability, 76–81
developmental hypotheses in the interpretation of Aristotle, 1, 7–8, 82, 105 n. 2, 220
deviant, 196–7, 208
dew, 94 n. 6, 116, 122
dialectic, 7, 15–16, 26, 34, 205, 208–9, 216–18, 221–2
diaphragm, 48
'diarrhoea', 89
dichotomists, Aristotle's criticism of, 55, 59, 71–2, 81, 108
diet, 108, 111, 191
differentiae, of animals, 49, 54–7, 60–1, 71, 81
differentiation, 190
  of parts of animals, 93, 100
digestion, 42, 47–8, 85–6, 89–90, 99–100, 102, 149, 154
dilemmatic arguments, 26
discrimination, 126, 129
discovery, 9, 102
disease, 45, 64, 86, 88–9
divine, the, 161
division, 97
dogfish, 93
'dualisers', 72 n. 5, 74–5, 77
dung, 94 n. 6, 116–17
dwarf-like, 120

ear, 128
earth, 45, 46 n. 23, 64, 70–1, 86, 93–4, 105, 120–1, 133, 149, 153, 171
earthquake, 163
'earth's guts', 115
eccentric, 167
eclipse, 101, 144, 168–9, 181
ecliptic, 169
ecology, 107
eels, 115
efficient causes, 28–9, 34, 42, 44, 51, 91–2, 95, 98–9, 105, 117–18, 121–4, 140, 146, 153
eggs, 90, 93–4, 100, 105, 108, 113, 115–6, 121
Egyptians, 162
elections, 187
elements, 63–4, 107, 139
elenchos, elenchtic demonstration, 20, 33, *see also* refutation
embryology, 85, 88, 91–3, 156
Empedocles, 26, 112–13, 124, 209–10
endoxa, 15, 18, 218
enmities between animals, 113, 186

enthymeme, 16–17
environment, 122, 124
epagōgē, 17, see also induction
epicycle, 167
epideictic oratory, 18, 215, 217
'epithets', 212
equality axiom, 143
equinoxes, precession of, 164
equivocation, 73, 81, 147
essence, 16, 20–2, 24, 27–8, 38, 52–3, 58, 60–1, 69, 72, 99–101, 115, 122, 143, 147–8
eternity, 27–9, 32, 34, 113, 122, 180–1
ethics, 18–19, 126, 203, 218–19
Euctemon, 177
Eudoxus, 160, 168, 172–3, 176–80
evaporation, 88, 152
evolution, 107, 113, 187
exactness, 13, 19, 21–2, 219
exchange, 193, 195
excluded middle, law of, 20, 72–3, 143
excreta, 84, 116
exhalations, 48, 166
experimentation, 115, see also thought experiment
eye, 40, 62 n. 79, 88–9, 93, 126, 128–9, 200
'eyes, before the', 212–13

'face' in the moon, 164–5
fat, 87, 91, 97, 108, 154
fear, 154
feathers, 71–2, 108–9, 141–2
female/male, 42, 50, 61, 91–4, 117–18, 120–4, 149, 151, 153, 185–6, 189–90, 198–200
fertilisation, 90–2, 102
fibre, 155
fig, 85
final causes, 29, 41, 44, 47, 51, 96, 99, 112, 122, 157, see also teleology
fire, 45, 46 n. 23, 64, 71, 85, 87, 94 n. 6, 116
fish, 55, 59, 71, 89–90, 105, 114, 121, 150
fishermen, 116, 192
fish-spine, 43, 132, 141, 145, 148–50
fixity of species, 113, 122
flame, 129
flesh, 'fleshy', 43–5, 51, 62, 76–8, 87–8, 116, 127–30, 132–3, 135–7, 141, 149, 153–4, 156
flying, 49, 57, 72, 80 n. 13, 108
focal meaning, 37, 139, 142–3, 207
foetus, 90, 100
food, 42, 50, 89, 98, 191–2, 194, see also nutrition
foot, 80 n. 13, 200

footless, 57
forensic oratory, 215, 217
form, 38–42, 44, 51–3, 61–3, 69–70, 84, 91–2, 95, 98–101, 105, 112, 118–19, 121–3, 139–40, 146, 153, 158, 190
  perceptible, 71, 126–9
foundations, 203
Frede, M., 53–4, 69
friendships between animals, 113, 186
frigidity of style, 212
front/back, 44, 48
fruit, 79, 85, 101
functionalism, 62, 65–6, 127 n. 1

gall-wasp, 86
generation, 42, 47, 51, 57, 79, 82, 91–5, 104, ch. 5, 149, 153
  spontaneous, 64 n. 81, 65, 77 n. 10, 91, 94–5, 100, ch. 5, 221
geometry, 27–8, 55–6, 143, 181, 216
gestation, 108, 111
glass, 70
gnats, 117
god, 112, 188
gold, 85
good, 139, 185, 218
Gotthelf, A., 8–9
grasshopper, 94
gregariousness, 186, 190
grey hair, 89, 151

habitat, 117, 134
hair, 43–4, 94 n. 6, 108–9, 116, 133, 152
haloes, 166
hand, 141, 149, 152
hard/soft, 128, 132, 134, 150, 153
harmonics, 143
hatching of eggs, 93–4, 100
health, 30–2, 45, 64, 84, 86, 105, 119 n. 21, 139–40, 196
hearing, 127–8, 131
heart, 27–8, 42 n. 5, 45 n. 16, 47–8, 50–1, 87, 98, 126–8, 131, 133, 136, 141, 149–50, 153, 155–6
heat/cold, 45–7, 64, 70, 83–97, 101, 116, 118, 120, 122–3, 128–9, 154, 156
  heat, extraneous/proper, 84–5, 89, 94–5, 119 n. 21, see also vital heat
heavenly bodies, 28, ch. 8
hen, 93–4, 121
herbivores, 191–2
herdsmen, 192
Herodotus, 112
Hesiod, 109, 112–13, 124
heuristic, 158

General index

hierarchy, 89, 108, 120, 123–5, 157, 171, 203
Hipparchus, 164
'hoar-frost', 151
holothouria, 75–6, 77 n. 10, 79, 81–2
Homer, 210, 216
homoeomerous/homogeneous substances, 63, 70, 97, 121, 149, *see also* uniform parts
homology, 152, 200
homonymy, 139
honour, 161
hoof, 141
horn, 70, 89, 108, 145, 151
horseflies, 117
household, 184–5, 191–4
Huainanzi, 107–111
humans, 56, 58, 60, 67, 108, 111, 127, 132, 171, 184–6, 192, 194, 202–3
   as model, 51, 81, 120, 123, 135–6, 187, 190, 200
Hume, 184–5, 201, 204
hunters, 192
hylozoism, 46, 65–6, 121–3
hypothesis, 12, 21, 102, 143
hypothetical syllogisms, 16

ichōr, see serum
ideals, 188, 197, 219–20
imagination, 155
indemonstrables, 11, 27–8, 34, 36, 100, 143, 209, 218
indeterminacy, 54, 68–9
induction, 17, 22, 218
insects, 41, 80 n. 13, 86, 106, 109, 114, 132, 153–4, 156–7, 191
instrumental parts, 41–2, 46, 51–2, 76
intelligence in animals, 44, 49, 51, 154, 186
intentionality, 104–5, 110
intermediates, 68, 73–5, 153
iron, 70

jellyfish (sea-lungs), 75–6, 77 n. 10, 79, 81–2
Jupiter, 162–3, 167, 171–3, 177

kidneys, 42, 47, 87, 98
kings, 190

land-animals, 49
lard, 45, 87–8, 91, 97
larvae (grubs), 80 n. 13, 94, 114, 117, 192
larvipara, 57, 190
law, 188, 201
law-suits, 187
leaves, 70, 101–2
leisure, 187

Lennox, J. G., 8
likeness, 208, 211, 218–19
lips, 42
literary criticism, 205, 214–15
litigants, 26
liver, 42, 45 n. 17, 87, 155
lobsters, 152
locomotion, 41, 47, 49–51, 57, 59, 72, 80 n. 13, 152–3, 156, *see also* movement
Lucretius, 113
lungs, 47, 152

magic, 110
magistracies, 187
marrow, 45, 87–8, 91, 97–8
Mars, 162, 171–3, 176–9
master/slave, 124, 189, *see also* slavery
mastering, 94
mastication, 86, 90
mathematics, 19, 33–4, 69, 101, 143, 159, 170–1, 173–5, 181–2, 219
matter, 22, 38–40, 42, 44–6, 51–6, 59–66, 68–71, 84, 87, 89, 91–2, 94–5, 98–9, 116–19, 121–4, 126, 139–40, 146–7, 152–3, 158–9
mean, 126–7, 133, 212
measure, 188
medium, 130–3, 136, 153
melt, 70, 129
membrane, 48, 88, 114, 130–1, 133–4
memory, 39–40
menses, 42 n. 7, 91–2, 149, 153
Mercury, 167 n. 9, 171–3, 177–9
mesentery, 42, 86
metals, 70
metamorphosis, 80 n. 13, ch. 5
metaphor, 37, 84–5, 138–9, 141, 147, 151, 157–8, 196, ch. 10
metaphysics, 10, 19–22, 118–19, 122, 138–9, 158, 168, 210, 219, 221
metics, 187
Meton, 177–8
milk, 84, 90, 92, 97, 192
Milky Way, 166
mind, 62–3
mining, 194
models, astronomical, 167–9, 172, 174, 176–82
mole, 81, 200
monarchy, 196
monsters, 105, 113, 120
moon, 144, 160, 162, 164–9, 171–3, 176, 178–9, 181
morality, 110, 186, 188, 202, *see also* ethics
'more and less', 57, 71–2, 140, 150

morphology, 57–8, 72, 152
mould, 151
mouth, 42, 50, 77, 86, 134
movement, 24, 53, 166, 169–70, 174, *see also* locomotion
mud, 94 n. 6, 105, 115–7, 122–4
mules, 26–7
mullet, 115
muscles, 43, 153
must, 84
mutilation, 115
mutis, 150, 155–6
myth, 108–9, 111–12, 124–5

nail, 43, 141
naked, 108–9, 111
nameless parts, 145, 149–50
natural philosophy/nature, 9, 23, 29–30, 69, 74, 99, 109–13, 121–2, 124–5, 146, 151, 158, ch. 9, 221
necessary parts, 49–50, 155
necessity, 12, 21–3, 29–34, 36–7, 47, 51, 118–19, 170, 174, 181–2, 190, 193–5
nerves, 50, 136
noble, 47 n. 26, 48
nomos, 185, 188–9, 198, 200–1, *see also* convention
normative, 120, 184, 196, 198, 200, 202, 208
nose, 196
novae, 164
numerology, 108
nutrition, 42, 44, 47–51, 57, 67, 86–7, 91–2, 98, 117, 154, 191

observation, 114–16, 161–2, 182
occultation, 162
octopus, 132 n. 3
oesophagus, 48
oil, 64, 87
old age, 45, 64, 88, 91, 93, 151, 216
oligarchy, 196
omentum, 42, 88, 97
order, 161, 164–5, 169, 171, 175, 180
organ, 58, 155, 158
  organ-formation, 100, 102
  see also instrumental parts, sense-organs
ostrich, 80 n. 13
ovipara, 57, 91, 93–4, 108, 190, 192
ovovivipara, 57
oyster, 77, 134

pain, 40
panther, 190
paradeigma, 17

paradox, 17
parboiling, 83, 85
Parmenides, 92
Peck, A. L., 154
Pellegrin, P., 54–5
perception, 16, 21–2, 40–4, 47–8, 50–1, 53, 56–8, 60, 67–8, 71–2, 74–82, ch. 6, 149, 153, 155–6, 161, 221
perfecting, 83–4, 101, 181
perspicuity, 211–12
persuasion, 17, 215, 217, 219, 222
phases, five, 107, 110
philosophy, practical, 205, 217–19
phlegm, 84
pig, 89, 108
pinna, 75, 77–8, 80
planets, 12–14, 162–82
plants, 41, 43, 46 n. 23, 48 n. 32, 56, 58, 64, 67–8, 71–80, 82, 85–6, 91, 94, 101–2, 114, 116, 119–21, 123, 125, 133, 149, 151, 157, 171, 185, 188, 192, 220–1
Plato, 3, 23–4, 65, 69, 79 n. 12, 161, 166, 169, 171–2, 175, 178, 180–1, 187, 196, 201, 209–10, 214 n. 10
Platonists, 182 n. 22
pleasure, 40, 161, 191, 211
Pliny, 164
pneuma, 46, 64, 120–2, 149
poetry, 112, 209–12, 215, 222
polis, city-state, 185–7
political constitutions, 195–6
politics, ch. 9, 218–19
polypod, 57
polythetic classification, 72, 81, 108
polyvalence, 37, 100–1, 103
poroi, 128
potentiality, *see* actuality
pounce, 132, 145, 148–9
predators, 113
principles, 8, 11, 18–24, 27–8, 34, 44, 48, 50, 139, 141–3, 157
privation, 60, 140, 146
probability, 200, 218
proportion, 84, 122, 138–9, 147, 149, 151–3, 157–8, 206, *see also* balance
Ptolemy, 164
purity, 45, 91, 153
purpuras, 115
pus, 84
putrefaction, 94, 105, 118, 151
Pythagoreans, 165

qi, 107–8, 110
quadruped, 41, 57, 80 n. 13, 81, 93–4, 200
qualitative change, 63, 126–7, 129

General index

ratio, 71
rawness, 83–4
razor-shells, 75, 78, 80
reason, 11, 39–40, 42 n. 6, 57, 67, 201
reductio, 13, 19, 22
refrigeration, 50
refutation, 17, 19–20, 26, 124, 208, *see also* elenchos
regularities, 105, 118, 120, 122, 124–5, 164–5, 169, 181, 198
rennet, 92
reproduction, 42, 50, 67, 91–5, 149, 155, 185–8
research, 114, 125, 170, 174, 182
resemblance of offspring to parents, 42 n. 7, 94
residue, 42, 47, 50, 62, 76–9, 81, 84, 86–8, 90–2, 94–5, 100, 116–17, 155–6
respiration, 50, 154
retroactive spheres, 160, 172–6
retrogradation, 164–5, 168–9, 176–7, 179–80
rhetoric, 10, 16–19, 24, 205, 209–10, 215–19, 222
right/left, 48, 156, 171, 190
ripening, 83–6, 101
rivals, 106, 124–5, 180, 209, 221
roasting, 83–5, 101
rock-plants, 77, 80
roots, 76, 78–9
rule/ruler, 50, 189, 193, 198, 201

sage, 108–9, 111
'said in many ways', 84, 96, 129
salt, 70
sameness, 138, 142, 146, 152, 157
sap, 13, 101–2
Saturn, 167, 170–3, 177
scale of being, 71, 94, 120, 124–5
scales (in fish), 71, 141–2
scaly animals, 108–9
scolopendra, 157
scorching, 83, 85
scorpions, 117
sea, 120, 149, 209
sea-anemones, 74–5, 80–1
sea-animals, 74
sea-lungs, *see* jellyfish
sea-nettles, 75
sea-squirts, 74–7, 80
sea-urchins, 90, 135, 137, 154
seal, 81
seasons, inequality of, 177–9
seed, 42, 79, 84, 86, 91, 101, 119–20, 155
selachians, 93, 150

self-preservation, 77–8
self-sufficiency, 185, 187
semantic stretch, 100, 103
semen, 47, 64, 91–2, 96–100, 102, 117, 120, 149, 153
sense-organs, 42–3, 51, 71, 76, 126, 128–31, 136, 156
sensorium, common, 126–7, 129, 131, 133, 155
serum, *ichōr*, 87–8, 155
sharks, 93
'shelled' animals, 108–9
sight, 71, 81, 126–9, 131
signs, 16
silk-worm, 114
simile, 206–7, 211, 216
Simplicius, 168, 176–8
sinew, 132
skin, 43, 88, 214
skin-winged, 152
slavery, 124, 184, 189–91, 193–4, 197–8, 200–1
sleep, 39, 45, 64, 129
smell, 71, 127, 131, 135
snails, 114, 134
snakes, 191
snow, 94 n. 6, 116, 122
social animals, 49, 186, 197
soda, 70
sophist, 209–10, 214 n. 10
Sorabji, R., 126–7
sound, 126, 131
speech, 42
spheres, 160, 172–81, *see also* concentric, retroactive
spines, of sea-urchins, 90
spleen, 87
sponges, 74–6, 80
stars, 46 n. 23, 149, 162, 166–7, 169, 171
stationary animals, 77 n. 10, 80 n. 13
stations of planets, 168
stomach, 42, 50, 86, 88–90, 92, 98–9, 155–6
stones, 70, 121
'strangeness' (in style), 206, 211–13
strict use of terms, 139, 207–8, 211, 213–14
struggle for existence, 113
style, 138, 207, 211–13, 215–17, 219
sublunary, 166, 169, 175, 180–1
substance, 39, 54–5, 58–61, 69–70, 84, 119, 141, 181
suet, 45, 87–8, 91, 97, 154
sun, 95, 116–19, 122, 124, 160, 166, 168–9, 171–3, 176–9
sunspots, 164–5

supernovae, 164
sweat, 88, 209
sweetness, 90
swimmers, 49, 57, 108
syllogistic, 7, 10–17, 24, 33, 35, 100, 142, 144, 167, 208–9, 221

tangibles, 131, 133 n. 4, 134
taste, 42, 71, 128, 134
teeth, 42, 50, 89, 108, 145, 155
teleology, 161, 171, 173, 180, *see also* final causes
tendons, 43
testacea, 58 n. 69, 74, 76–8, 80–1, 114, 117, 120, 123, 132–5, 149, 151, 154, 155 n. 9, 156, 214
Thales, 121–2
'theologians', 112, 124
theology, 22
theoretical sciences, 29–32, 34
Thompson, D'A. W., 76
thought experiment, 131, 133
thunder, 143–4
tongue, 42, 50, 134, 155
top-down explanations, 64, 99 n. 10
Torone, 76
touch, 43–4, 50, 57, 58 n. 69, 68, 71, 80, 126–36, 149, 153
transfer, 85, 205–6, 209, 211, 214, *see also* metaphor
truth, 19
tumours, 84
twinkling, 13–14, 33, 167
tyranny, 196

unchanging, 164, 180–1
unclarity, the unclear, 25, 166, 209, 211–12, 218
understanding, 8, 11, 141–2
uniform parts, 44–6, 51–2, 61, 64, *see also* homoeomerous/homogeneous substances
univocity, 35, 37, 100–3, 208
unmoved movers, 174, 178, 180
up/down, 48, 120, 156, 169, 190
urine, 70, 84
useful versus useless residues, 86

validity, 100, 141, 208
value, value judgements, 46 n. 26, 48, 120, 201–2
Venus, 167, 171–3, 176–9
viscera, 44 n. 14, 45 n. 17, 48, 87–8, 154
vital functions, 40, 45, 50–2, 56–9, 62, 72, 82, 86, 99, 101, 106, 116, 118, 149, 156–7
vital heat, 46, 52, 64–5, 133–4
vitalist models, 63, 65
vividness, 212
vivipara, 57, 91, 93–4, 98, 100, 108, 150–1, 191–2
voice, 42

walking, 57, 72–3
'wandering' of planets, 169, 175
Wang Chong, 107, 110
warfare, 192–4
water, 45, 46 n. 23, 64, 70–1, 85, 120–1
ways of life, 49, 186, 191
wealth, 191–5, 197
weather, 93–5, 116–17, 122
wet/dry, 45–6, 52, 64, 83–6, 88–9, 95–7, 101, 128, 134, 153
whey, 70
white/black, 126
wild, 111, 192
wind, 94 n. 5
'wind-eggs', 90, 93
windpipe, 48
wine, 70
wing, 41, 71–2, 114, 152, 200
wisdom, 112 n. 9, 149, 210
wit, 213
womb, 92–3
wonder, 104, 125
wood, 85, 94 n. 6, 116–17, 122
'wood-carrier', 114–16
woody, 152
wool, 96 n. 6, 116

yin yang, 107
yolk, 93
young/old, 88, 91 n. 4, 189, 198

zodiac, 165

For EU product safety concerns, contact us at Calle de José Abascal, 56–1°,
28003 Madrid, Spain or eugpsr@cambridge.org.